AF564911

BASICS OF ENVIRONMENTAL CHEMISTRY

BASICS OF ENVIRONMENTAL CHEMISTRY

C.K. Choudhary

ANMOL PUBLICATIONS PVT. LTD.
NEW DELHI - 110 002 (INDIA)

ANMOL PUBLICATIONS PVT. LTD.
H.O.: 4374/4B, Ansari Road, Daryaganj,
New Delhi-110 002 (India)
Ph.: 23278000, 23261597
B.O.: No. 1015, Ist Main Road, BSK IIIrd Stage
IIIrd Phase, IIIrd Block,
Bangalore - 560 085 (India)
Visit us at: www.anmolpublications.com

Basics of Environmental Chemistry

First Edition, 2009
ISBN 978-81-261-3774-9

PRINTED IN INDIA

Printed at Balaji Offset Press, Delhi.

Contents

Preface

Environmental Chemistry is the scientific study of the chemical and biochemical phenomena that occur in natural places. It should not be confused with green chemistry, which seeks to reduce potential pollution at source. It can be defined as the study of the sources, reactions, transport, effects, and fates of chemical species in the air, soil, and water environments; and the effect of human activity on these. Environmental chemistry is an interdisciplinary science that includes atmospheric, aquatic and soil chemistry, as well as heavily relying on analytical chemistry and being related to environmental and other areas of science.Environmental chemists draw on a range of concepts from chemistry and various environmental sciences to assist in their study of what is happening to a chemical species in the environment. Important general concepts from chemistry include understanding chemical reactions and equations, solutions, units, sampling, and analytical techniques

The two components of nature, the organism and the environment, are not only much complex and dynamic but also interdependent, mutually reactive and interrelated. So, environment is very much important for our sustainable development.Therefore, a new environmental behaviour and management is necessary to cope with various trenchant challenges of the environment, in which quantitative demands and confrontation must be replaced by qualitative appreciation and co-ordination. This will certainly lead us into a new era of harmony where all care for the well being of the life on watery planet.

Author

1

Pollution

INTRODUCTION

Pollution is the introduction by man into the environ-ment of contaminants that cause harm or discomfort to humans or other living organisms, or damage the environment

Pollution can be in the form of chemical substances, or energy such as noise, heat, or light. Pollutants can be naturally occurring substances or energies, but are considered contaminants when in excess of natural levels. Pollution is often categorized into point source and no point source pollution.An important aspect of the notion of pollution is that ecological change must actually be demonstrated. If some potentially polluting substance is present at a concentration or intensity that is less than the threshold required to cause a demonstrable ecological change, then the situation would be referred to as contamination, rather than pollution.

Some other elements can be present in very large concentrations, aluminum and iron, which are important constituents of rock and soil. Aluminum constitutes 8-10% of the earth's crust and iron 3-4%. However, almost all of the aluminum and iron present in minerals are insoluble in water and are therefore not readily assimilated by biotic community and cannot cause toxicity. In acidic environments, however, ionic forms of aluminum are solubilized, and these can cause toxicity in concentrations of less than one part per million. Therefore, the bio-availability of a chemical is an important determinant of whether its presence in some concentration will cause pollution.

Most instances of pollution result from the activities of humans can be caused by:

- The emission of sulfur dioxide and metals from a smelter, causing toxicity to vegetation and acidifying surface waters and soil,
- The emission of waste heat from an electricity generating station into a river or lake, causing community change through thermal stress, or
- The discharge of nutrient-containing sewage wastes into a water body, causing eutrophication.

Most instances of anthropogenic pollution have natural analogues, that is, cases where pollution is not the result of human activities.Pollution can be caused by the emission of sulfur dioxide from volcanoes, by the presence of toxic elements in certain types of soil, by thermal springs or vents, and by other natural phenomena. In many cases, natural pollution can cause an intensity of ecological damage that is as severe as anything caused by anthropogenic pollution.

An interesting case of natural air pollution is the Smoking Hills, located in a remote and pristine wilderness, virtually uninfluenced by humans. However, at a number of places along the 18.63 miles (30 km) of seacoast, bituminous shales in sea cliffs have spontaneously ignited, causing a fumigation of the tundra with sulfur dioxide and other pollutants. The largest concentrations of sulfur dioxide (more than two parts per million) occur closest to the combustions. Further away from the sea cliffs the concentrations of sulfur dioxide decrease rapidly. The most-important chemical effects of the air pollution are acidification of soil and fresh water, which in turn causes a solubilization of toxic metals. Surface soils and pond waters commonly have pHs less than 3, compared with about Ph 7 at non-fumigated places. The only reports of similarly acidic water are for volcanic lakes in Japan, in which natural pHs as acidic as 1 occur, and pH less than 2 in waters affected by drainage from coal mines.

At the Smoking Hills, toxicity by sulfur dioxide, acidity, and water-soluble metals has caused great damage to ecological communities. The most- intensively fumigated terrestrial sites have no vegetation, but further away a few pollution-tolerant species are present. About one kilometer away the toxic stresses are low enough that reference tundra is present. There are a few pollution-tolerant algae in the acidic ponds, with a depauperate community of six species occurring in the most-acidic pond in the area.Other cases of natural pollution concern places where certain elements are present in toxic amounts. Surface mineralizations can have toxic metals present in large concentrations,copper at 10% in peat at a copper-rich spring in New Brunswick, or surface soil with 3% lead plus zinc on Baffin Island. Soils influenced by nickel-rich serpentine minerals have been well-studied by ecologists. The stress-adapted plants of serpentine habitats form distinct communities, and some plants can have nickel concentrations larger than 10% in their tissues. Similarly, natural soils with large concentrations of selenium support plants that can hyperaccumulate this element to concentrations greater than 1%. These plants are poisonous to livestock, causing a toxic syndrome known as *blind staggers*.

There are many well-known cases where pollution is caused by anthropogenic emissions of chemicals. Some examples include:

- Emissions of sulfur dioxide and metals from smelters can cause damage to surrounding terrestrial and aquatic ecosystems. The sulfur dioxide and metals are directly toxic. In addition, the deposition of sulfur dioxide can cause an extreme acidification of soil and water, which causes metals to be more bio-available, resulting in important, secondary toxicity. Because smelters are point sources of emission, the spatial pattern of chemical pollution and ecological damage displays an exponentially decreasing intensity with increasing distance from the source.
- The use of pesticides in agriculture, forestry, and around homes can result in a non-target exposure of birds and

other wildlife to these chemicals. If the non-target biota is vulnerable to the pesticide, then ecological damage will result. During the 1960s urban elm trees in the eastern United States were sprayed with large quantities of the insecticide DDT, in order to kill beetles that were responsible for the transmission of Dutch elm disease, an important pathogen. Because of the very large spray rates, many birds were killed, leading to reduced populations in some areas. Birds and other non-target biota have also been killed by modern insecticide-spray progammes in agriculture and in forestry.

- The deposition of acidifying substances from the atmosphere, mostly as acidic precipitation and the dry deposition of sulfur dioxide can cause an acidification of surface waters. The acidity solubilizes metals, most notably aluminum, making them bio-available. The acidity in combination with the metals causes toxicity to the biota, resulting in large changes in ecological communities and processes. Fish are highly intolerant of acidic waters.
- Oil spills from tankers and pipelines can cause great ecological damage. When oil spilled at sea washes up onto coastlines, it destroys seaweeds, invertebrates, and fish, and their communities are changed for many years. Seabirds are very intolerant of oil and can die of hypothermia if even a small area of their feathers is coated by petroleum.
- Most of the lead shot fired by hunters and skeet shooters miss their target and are dispersed into the environment. Waterfowl and other avian wildlife actively ingest lead shot because it is similar in size and hardness to the grit that they ingest to aid in the mechanical abrasion of hard seeds in their gizzard.

Humans can also cause pollution by excessively fertilizing natural ecosystems with nutrients.Freshwaters can be made eutrophic by fertilization with phosphorus in the form of phosphate. The most conspicuous symptoms of eutrophication are changes in species composition of the phytoplankton community and, especially, a large increase in algal biomass, known as a *bloom*. In shallow water bodies there may also be a vigourous growth of vascular plants. These primary responses are usually accompanied by secondary changes at higher trophic levels, including arthropods, fish, and waterfowl, in response to greater food availability and other habitat changes. However, in the extreme cases of very eutrophic waters, the blooms of algae and other microorganisms can be noxious, producing toxic chemicals and causing periods of oxygen depletion that kill fish and other biota. Extremely eutrophic water bodies are polluted because they often cannot support a fishery, cannot be used for drinking water, and have few recreational opportunities and poor esthetics.

Current Protocols

Protocols exist for avoiding pollution of other worlds. Payloads are scrupulously sterilized. Rocket stages that loft probes toward these targets are diverted to avoid trajectories that would follow the probe.Pollution is a harmful change in the natural environment caused by human activities:This may be the release of substances which are toxic to either animals or plants, or it may be the release of energy (heat, light, radiation, or sound) which interferes with the development of animals or plants.Pollution is usually harmful to one or more species of animal or plant. However, releasing nutrients into rivers may be beneficial to some organisms while it is harmful to others.

Here are some general examples:

- Automobile exhaust emissions
- Oil spills
- The dumping of industrial wastes in the water supply
- The overuse of pesticides
- The overuse of chemical fertilizers
- Improper disposal of solid wastes

Releasing raw sewage into a river causes pollution. Sewage contains lots of organic substances which are a source of food for bacteria and fungi. The result is that the bacteria and fungi thrive. Since they use up all the oxygen in the water, most of the animals will die.Pollution may occur naturally, as when an erupting volcano emits sulphur dioxide, but the term usually refers to the negative effect of human activities

POLLUTION PREVENTION

Pollution prevention is reducing or eliminating waste at the source by modifying production processes, promoting the use of non-toxic or less-toxic substances, implementing conservation techniques, and re-using materials rather than putting them into the waste stream. Pollution prevention means "source reduction," as defined under the Pollution Prevention Act, and other practices that reduce or eliminate the creation of pollutants through:

- Increased efficiency in the use of raw materials, energy, water, or other resources, or
- Protection of natural resources by conservation.

The Pollution Prevention Act defines "source reduction" to mean any practice which:

- Reduces the amount of any hazardous substance, pollutant, or contaminant entering any waste stream or otherwise released into the environment (including fugitive emissions) prior to recycling, treatment, or disposal; and
- Reduces the hazards to public health and the environment associated with the release of such substances, pollutants, or contaminants.

The term includes: equipment or technology modifications, process or procedure modifications, reformulation or redesign or products, substitution of raw materials, and improvements in housekeeping, maintenance, training, or inventory control.

Specific Pollution Prevention Approaches

Pollution prevention approaches can be applied to all pollution-generating activities, including those found in the energy, agriculture, Federal, consumer, as well as industrial sectors. The impairment of wetlands, ground water sources, and other critical resources constitutes pollution, and prevention practices may be essential for preserving these resources. These practices may include conservation techniques and changes in management practices to prevent harm to sensitive ecosystems. Pollution prevention does not include practices that create new risks of concern.

In the agricultural sector, pollution prevention approaches include:

- Reducing the use of water and chemical inputs;
- Adoption of less environmentally harmful pesticides or cultivation of crop strains with natural resistance to pests; and
- Protection of sensitive areas.

In the energy sector, pollution prevention can reduce environmental damages from extraction, processing, transport, and combustion of fuels. Pollution prevention approaches include:

- Increasing efficiency in energy use;
- Substituting environmentally benign fuel sources; and
- Design changes that reduce the demand for energy.

Some Types of Pollution are: Air,Water and thermal, Radioactive.

AIR POLLUTION

One of the formal definitions of air pollution is as follows— 'The presence in the atmosphere of one or more contaminants in such quality and for such duration as is injurious, or tends to be injurious, to human health or welfare, animal or plant life.' It is the contamination of air by the discharge of harmful substances. Air pollution can cause health problems and it can also damage the environment and property. It has caused thinning of the protective ozone layer of the atmosphere, which is leading to climate change. Modernisation and progress have led to air getting more and more polluted over the years. Industries, vehicles, increase in the population, and urbanization are some of the major factors responsible for air pollution. The following industries are among those that emit a great deal of pollutants into the air: thermal power plants, cement, steel, refineries, petro chemicals, and mines. Air pollution results from a variety of causes, not all of which are within human control. Dust storms in desert areas and smoke from forest fires and grass fires contribute to chemical and particulate pollution of the air. The source of pollution may be in one country but the impact of pollution may be felt elsewhere. The discovery of pesticides in Antarctica, where they have never been used, suggests the extent to which aerial transport can carry pollutants from one place to another.

Sources of Air Pollution

The combustion of gasoline and other hydrocarbon fuels in automobiles, trucks, and jet airplanes produces several primary pollutants: nitrogen oxides, gaseous hydrocarbons, and carbon monoxide, as well as large quantities of particulates, chiefly lead. In the presence of sunlight, nitrogen oxides combine with hydrocarbons to form a secondary class of pollutants, the photochemical oxidants, among them ozone and the eye-stinging peroxyacetylnitrate (PAN). Nitrogen oxides also react with oxygen in the air to form nitrogen dioxide, a foul-smelling brown gas. In urban areas like Los Angeles where transportation is the main cause of air pollution, nitrogen dioxide tints the air, blending with other contaminants and the atmospheric water vapour to produce brown smog. Although the use of catalytic converters has reduced smog-producing compounds in motor vehicle exhaust emissions, recent studies have shown that in so doing the converters produce nitrous oxide, which contributes substantially to global warming.

Air may be severely polluted not only by transportation but also by the burning of fossil fuels (oil and coal) in generating stations, factories, office buildings, and homes and by the incineration of garbage. The massive combustion produces tons of ash, soot, and other particulates responsible for the gray smog of cities like New York and Chicago, along with enormous quantities of sulfur oxides (which also may be result from burning coal and oil). These oxides rust iron, damage building stone, decompose nylon, tarnish silver, and kill plants. Air pollution from cities also affects rural areas for many miles downwind.Every industrial process exhibits its own pattern of air pollution. Petroleum refineries are responsible for extensive hydrocarbon and particulate pollution. Iron and steel mills, metal smelters, pulp and paper mills, chemical plants, cement and asphalt plants—all discharge vast amounts of various particulates. Uninsulated high-voltage power lines ionize the adjacent air, forming ozone and other hazardous pollutants. Airborne pollutants from other sources include insecticides, herbicides, radioactive fallout, and dust from fertilizers, mining operations, and livestock feedlots.

Sources And Major Air Ppollutants

Carbon monoxide (CO) is a colourless, odourless gas that is produced by the incomplete burning of carbon-based fuels including petrol, diesel, and wood. It is also produced from the combustion of natural and synthetic products such as cigarettes. It lowers the amount of oxygen that enters our blood. It can slow our reflexes and make us confused and sleepy.

Carbon dioxide (CO_2) is the principle greenhouse gas emitted as a result of human activities such as the burning of coal, oil, and natural gases.

Chloroflorocarbons (CFC) are gases that are released mainly from air-conditioning systems and refrigeration. When released into the air, CFCs rise to the stratosphere, where they come in contact with few other gases, which leads to a reduction of the ozone layer that protects the earth from the harmful ultraviolet rays of the sun.

Lead is present in petrol, diesel, lead batteries, paints, hair dye products, etc. Lead affects children in particular. It can cause nervous system damage and digestive problems and, in some cases, cause cancer.

Ozone occur naturally in the upper layers of the atmosphere. This important gas shields the earth from the harmful ultraviolet rays of the sun. However, at the ground level, it is a pollutant with highly toxic effects. Vehicles and industries are the major source of ground-level ozone emissions. Ozone makes our eyes itch, burn, and water. It lowers our resistance to colds and pneumonia.

Nitrogen oxide (Nox) causes smog and acid rain. It is produced from burning fuels including petrol, diesel, and coal. Nitrogen oxides can make children susceptible to respiratory diseases in winters.

Suspended particulate matter (SPM) consists of solids in the air in the form of smoke, dust, and vapour that can remain suspended for extended periods and is also the main source of haze which reduces visibility. The finer of these particles, when breathed in can lodge in our lungs and cause lung damage and respiratory problems.

Sulphur dioxide (SO_2) is a gas produced from burning coal, mainly in thermal power plants. Some industrial processes, such as production of paper and smelting of metals, produce sulphur dioxide. It is a major contributor to smog and acid rain. Sulfur dioxide can lead to lung diseases.

Chemical Pollution

In some parts of the world, the bodies of whales and dolphins washing ashore are so highly contaminated that they qualify as toxic waste and have to be specially disposed of. There are many different sources of chemical pollution, including:

- Domestic sewage
- Industrial discharges
- Seepage from waste sites
- Atmospheric fallout
- Domestic run-off
- Accidents and spills at sea
- Operational discharges from oil rigs
- Mining discharges and
- Agricultural run-off.

Health Effects of Air Pollution

The human health effects of poor air quality are far reaching, but principally affect the body's respiratory system and the cardiovascular system. Individual reactions to air pollutants depend on the type of pollutant a person is exposed to, the degree of exposure, the individual's health status and genetics. People who exercise outdoors on hot smoggy days increase their exposure to pollutants in the air.The health effects caused by air pollutants may range from subtle biochemical and physiological changes to difficulty breathing, wheezing, coughing and aggravation of existing respiratory and cardiac conditions. These effects can result in increased medication use, increased doctor or emergency room visits, more hospital admissions and even premature death.

- Human Respiratory System
- Human Cardiovascular System
- Heart and Lung Diseases
- Pyramid of Health Effects
- Populations at Risk
- Leading Causes of Hospitalization
- Leading Causes of Death
- Estimating Health Benefits

HUMAN RESPIRATORY SYSTEM

The health of our lungs and entire respiratory system is affected by the quality of the air we breathe. In addition to oxygen, this air contains other substances such as pollutants, which can be harmful. Exposure to chemicals by inhalation can negatively affect our lungs and other organs in the body. The respiratory system is particularly sensitive to air pollutants because much of it is made up of exposed membrane. Lungs are anatomically structured to bring large quantities of air (on average, 400 million litres in a lifetime) into intimate contact with the blood system, to facilitate the delivery of oxygen.

Lung tissue cells can be injured directly by air pollutants such as ozone, metals and free radicals. Ozone can damage the alveoli — the individual air sacs in the lung where oxygen and carbon dioxide are exchanged. More specifically, airway tissues which are rich in bioactivation enzymes can transform organic pollutants into reactive metabolites and cause secondary lung injury. Lung tissue has an abundant blood supply that can carry toxic substances and their metabolites to distant organs. In response to toxic insult, lung cells also release a variety of potent chemical mediators that may critically affect the function of other organs such as those of the cardiovascular system. This response may also cause lung inflammation and impair lung function.

Structure and Function

The human respiratory system is dominated by our lungs, which bring fresh oxygen (O_2) into our bodies while expelling carbon dioxide (CO_2). The oxygen travels from the lungs through the bloodstream to the cells in all parts of the body. The cells use the oxygen as fuel and give off carbon dioxide as a waste gas. The waste gas is carried by the bloodstream back to the lungs to be exhaled.The lungs accomplish this vital process - called gas exchange - using an automatic and quickly adjusting control system. This gas exchange process occurs in conjunction with the central nervous system (CNS), the circulatory system, and the musculature of the diaphragm and the chest.

The human respiratory system can be divided into the upper respiratory tract and the lower respiratory tract. The upper respiratory tract includes the following rigid structures:

Nasal cavities: Filter the air we breathe and provide a sense of smell.

Pharynx: Acts in the respiratory and the digestive system.

Larynx: Link between the pharynx and the trachea. Generates the voice with the presence of vocal folds.

Trachea: The trachea is the bond with the lower respiratory tract. This is a flexible structure allowing the air to go down to the lungs.

The lungs and the other parts of the respiratory system have important jobs to do related to breathing. These include:

- Bringing all air to the proper body temperature.
- Moisturizing the inhaled air for necessary humidity.
- Protecting the body from harmful substances by coughing, sneezing, filtering or swallowing them, or by alerting the body through the sense of smell.

Defending the lungs with cilia (tiny hair-like structure), mucus and macrophages, which act to remove harmful substances deposited in the respiratory system.

Causes of Air Pollution

Humans probably first experienced harm from air pollution when they built fires in poorly ventilated caves. Since then we have gone on to pollute more of the earth's surface. Until recently, environmental pollution problems have been local and minor because of the Earth's own ability to absorb and purify minor quantities of pollutants. The industrialization of society, the introduction of motorized vehicles, and the explosion of the population, are factors contributing toward the growing air pollution problem. At this time it is urgent that we find methods to clean up the air.

The primary air pollutants found in most urban areas are carbon monoxide, nitrogen oxides, sulfur oxides, hydrocarbons, and particulate matter (both solid and liquid). These pollutants are dispersed throughout the world's atmosphere in concentrations high enough to gradually cause serious health problems. Serious health problems can occur quickly when air pollutants are concentrated, such as when massive injections of sulfur dioxide and suspended particulate matter are emitted by a large volcanic eruption.

Air Pollution in the Home

You cannot escape air pollution, not even in your own home. "The Environmental Protection Agency (EPA) reported that toxic chemicals found in the air of almost every American home are three times more likely to cause some type of cancer than outdoor air pollutants". The health problems in these buildings are called "sick building syndrome". The EPA has found that the air in some office buildings is 100 times more polluted than the air outside. Poor ventilation causes about half of the indoor air pollution problems. The rest come from specific sources such as copying machines, electrical and telephone cables, mold and microbe-harboring air conditioning systems and ducts, cleaning fluids, cigarette smoke, carpet, latex caulk and paint, vinyl molding, linoleum tile, and building materials and furniture that emit air pollutants such as formaldehyde. A major indoor air pollutant is radon-222, a colourless, odorless, tasteless, naturally occurring radioactive gas produced by the radioactive decay of uranium-238. "According to studies by the EPA and the National Research Council, exposure to radon is second only to smoking as a cause of lung cancer".

WATER

Water is life ! It is a precondition for human, animal and plant life as well as an indispensable resource for the economy. Water also plays a fundamental role in the climate regulation cycle.Protection of water resources, of fresh and salt water ecosystems and of the water we drink and bathe in is therefore one of the cornerstones of environmental protection in Europe. The stakes are high and the issues transcend national boundaries and concerted action at the level of the EU is necessary to ensure an effective protection.

WASTE AND WATER POLLUTION

When toxic substances enter lakes, streams, rivers, oceans, and other water bodies, they get dissolved or lie suspended in water or get deposited on the bed. This results in the pollution of water whereby the quality of the water deteriorates, affecting aquatic ecosystems. Pollutants can also seep down and affect the groundwater deposits.The effects of water pollution are not only devastating to people but also to animals, fish, and birds. Polluted water is unsuitable for drinking, recreation, agriculture, and industry. It diminishes the aesthetic quality of lakes and rivers. More seriously, contaminated water destroys aquatic life and reduces its reproductive ability. Eventually, it is a hazard to human health. Nobody can escape the effects of water pollution.

Importance of Water

Water is our lifeline that bathes us and feeds us. In ancient cultures water represented the very essence of life. The Romans were the first to pipe water into their growing cities, especially with their aqueducts. They also realised that sewage water could cause damage to their people, and needed to be removed from large areas of people.

Water has played a role not only in the history of countries, but in religion, mythology, and art. Water in many religions cleanses the soul through holy water.The water at Lourdes, France is thought by many religions to be sacred water with healing powers. In Egyptian mythology, the Nu was the beginning of everything and represented water. It brought life to their people, but in drought, produced chaos.

The water or hydrologic cycle explains interactions between the atmosphere, hydrosphere, and lithosphere. The water or hydrologic cycle is a major driving force on our planet. Water is in constant motion, evapourating into the atmosphere from oceans, lakes, rivers and streams. When the atmosphere can no longer support the moisture within the clouds, we experience rain, snow, hail, or sleet. Some water is locked in the form of ice at the polar caps and in glaciers. Water melts in the spring, producing runoff, that percolates through the Earth as groundwater (subsurface) or makes its way back to the sea (surface). The oceans contain most of the water, but it is salt water which is unusable by most organisms. Only pure H_2O (water) can interact with organisms.

The movement of the oceans also has a direct effect on the atmosphere. The atmosphere is that envelope of gas that keeps organisms living on this planet. Oceans and atmosphere interact to give us weather.

Water provides the Earth with the capacity of supporting life. An organism doesn't have to be told how important water is to their existence. An amphibian knows to lay their eggs in water or else there will be no new born. Even flies know to lay their eggs in fresh water.The only organism that doesn't understand the importance of water is humans, especially in industrialized countries. Children in those societies turn on the water in a sink and never think about the trouble someone has gone for that "miracle" to occur.

Dams, reservoirs, filtering plants, and pipes all bring clean water when the facet is turned on. Sewage water is only mixed with recycled water supplies after the water goes through rigorous cleaning methods. Water borne diseases are any illnesses caused by drinking contaminated water. Diseases can include infection from bacteria (Salmonella), viruses, or by small parasites (Cryptosporida, Giardia, and Toxoplasma). These organisms and viruses cause diseases like cholera, typhoid fever, malaria, botulism, polio, dysentery, giardia, and hepatitis A. One of the first symptoms of these diseases is diarrhea, which cause about three million deaths throughout the world.Sewage is sometimes discharged into rivers, where children downstream might be taking a bath or using the water to drink. The simplest treatment method is boiling. Just bring the water to a boil for at least one minute, then allow it to cool. But this is not always effective in heavily chemical polluted water supplies.

Without water, organisms could not exist. Water is a resource that should not be taken for granted. It needs to be conserved, just as we save other valuable resources.

Water is one of the weirdest compounds known to humans. The difference between the boiling point and freezing point of water is one of the largest ranges of any compound. It is this span of temperature that mirrors the range of where life can exist, from bacteria to humans. Water also has a very high specific heat, which means that it can absorb or lose much heat before its temperature changes. This is important in maintaining body heat in mammals. It also takes a lot of energy before vapourization can occur. For this reason, water evapourates slowly from ponds and lakes, where many life forms are dependent on a stable, warm environment.

Water is less dense in its solid state than in its liquid state, so that ice floats instead of sinking. This property permits life to develop in polar and subpolar regions where ice floats and allows life to continue living below the surface. If ice were heavier than water, it would sink, and more ice would form on top of it. As a result, all life in the waters would be trapped in the ice in the many areas of the world where it gets cold enough to freeze water.

Resistance of water to a disturbance.

Water also exhibits viscosity. One can observe the effects of viscosity alongside a stream or river with uniform banks. The water along the banks is nearly still, while the current in the centre may be swift. This resistance between the layers is called viscosity. This property allows smaller fish to live near the shore, while larger fish are able to swim efficiently in strong currents. Viscosity is also responsible for the formation of eddies, creating turbulence that leads to good mixing of air in the water and more uniform distribution of microscopic organisms.

Domestic sewage refers to waste water that is discarded from households. Also referred to as sanitary sewage, such water contains a wide variety of dissolved and suspended impurities. It amounts to a very small fraction of the sewage by weight. But it is large by volume and contains impurities such as organic materials and plant nutrients that tend to rot. The main organic materials are food and vegetable waste, plant nutrient come from chemical soaps, washing powders, etc. Domestic sewage is also very likely to contain disease-causing microbes. Thus, disposal of domestic waste water is a significant technical problem.

Industrial effluents

Waste water from manufacturing or chemical processes in industries contributes to water pollution. Industrial waste water usually contains specific and readily identifiable chemical compounds. During the last fifty years, the number of industries in India has grown rapidly. But water pollution is concentrated within a few subsectors, mainly in the form of toxic wastes and organic pollutants. Out of this a large portion can be traced to the processing of industrial chemicals and to the food products industry. In fact, a number of large- and medium-sized industries in the region covered by the Ganga Action Plan do not have adequate effluent treatment facilities. Most of these defaulting industries are sugar mills, distilleries, leather processing industries, and thermal power stations. Most major industries have treatment facilities for industrial effluents. But this is not the case with small-scale industries, which cannot afford enormous investments in pollution control equipment as their profit margin is very slender.

Effects of water pollution

The effects of water pollution are not only devastating to people but also to animals, fish, and birds. Polluted water is unsuitable for drinking, recreation, agriculture, and industry. It diminishes the aesthetic quality of lakes and rivers. More seriously, contaminated water destroys aquatic life and reduces its reproductive ability. Eventually, it is a hazard to human health. Nobody can escape the effects of water pollution.

Effects of Run-off Pollution

Rain picks up dirt and silt and carries it into the water. If the dirt and silt settle in the water body, then these sediments prevent sunlight from reaching aquatic plants. If the Sun can't reach the plants, these perish. These sediments also clog fish gills and smother organisms that live on the bottom of the body of the water.

Effects of Oil Pollution and Antifreeze

If oil is spilled on the water, the effects on the ecosystem and the components are harmful. Many animals can be annihilated in case they ingest oil. Oil contaminated prey may be a reason of death for many. If the oil coats the feathers of birds, these may die. Oil and antifreeze makes the water have a foul odor and there is a sticky film on the surface of water that kills animals. Oil is the most harmful pollutant in the water.

Contaminated Ground Water Effects

If contaminated water enters the ground, there may be serious effects. People may become very sick and there is a probability of developing liver or kidney problems and cancer or other illnesses.

Fertilizers and other chemicals

Nitrates in drinking water leads to diseases of infants that may lead to their death. Cadmium is a metal in sludge-derived fertilizer. This can be absorbed by crops. When people ingest this, they may cause diarrheal disorders, liver and kidney damage. The inorganic substances like mercury, arsenic and lead are the causes of pollution. Other chemicals can also lead to problems concerning the taste, smell and colour of water. Pesticides, PCBs and PCPs are all poisonous to all sorts of life. Pesticides are used in farming, homes and forestry. PCBs are found as insulators in old electrical transformers. PCPs are found in products like wood preservatives.

Effects of Agricultural Water Pollution

Rain and irrigation water drains off cultivated land that has been fertilized and treated with pesticides, the excess nitrogen and poisons are mixed with it into the water supply. These pesticides are toxic and pollute the water in a different mode. Aquatic plants growth cause deoxygenation of water and annihilate flora and fauna in a stream, lake and river. Fertilizers enhance the growth of bacteria that are in water and increase the concentration of bacteria to hazardous levels.

Effects of Thermal Water Pollution

Machinery in the industries are cooled with water from lakes and rivers. This water reaches the river in a heated state. This water decreases the ability of the aquatic system to hold oxygen and raises the growth of warm water species.

Effects of Heavy Metal Water Pollution

Heavy metals like lead, mercury, iron, cadmium, aluminum and magnesium are present in water sources. If these metals are present in the sediment, these reach the food chain through plants and aquatic animals. This causes heavy metal poisoning in case the level in the water is very high.

THERMAL POLLUTION

Thermal pollution describes the introduction of waste heat into water bodies such as rivers, estuaries and coastal waters. Although other industries abstract water and then discharge it, the coastal power plants produce the greatest amount of water and thus waste heat. Coastal power stations of up to 1000MW capacity discharge up to 50cumecs ($m^3 s^{-1}$) at elevated temperatures. The effects of that discharge depend on the time and place of discharge and the characteristics of the receiving area. In particular, the effects depend on the ability to disperse and absorb the additional energy.

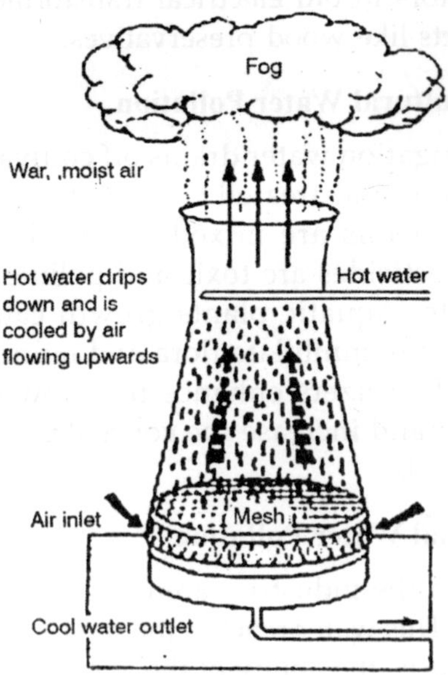

Fig:Thermal Power Plant

Power stations with a ready supply of cold water, such as those on the coast and in estuaries, use this to condense the steam used to turn the turbines. This flow of direct cooling, as opposed to the use of cooling towers, returns water to an adjacent area up to 10 degrees centigrade above ambient. The fossil and nuclear fuel power plants require this cooling. The high cost-effective value of direct cooling explains the large number of directly cooled estuarine and coastal power plants worldwide.

ECOSYSTEMS CAUSED BY HEAT (THERMAL POLLUTION)

There are various effects on the biology of the ecosystems when heated effluents reach the receiving waters. The species that are intolerant to warm conditions may disappear, while others, rare in unheated water, may thrive so that the structure of the community changes. Thermal pollution can have a great influence on the aquatic ecosystem. Species that are restricted to heated waters, can build up large populations in the receiving waters. Respiration and growth rates may be changed and these may alter the feeding rates of organisms. The reproduction period may be brought forward and development may be speeded up. Parasites and diseases may also be affected.

An increase of temperature also means a decrease in oxygen solubility. Any reduction in the oxygen concentration of the water, particularly when organic pollution is also present, may result in the loss of sensitive species.

Major Sources

The major sources of thermal pollution are electric power plants and industrial factories. In most electric power plants, heat is produced when coal, oil, or natural gas is burned or nuclear fuels undergo fission to release huge amounts of energy. This heat turns water to steam, which in turn spins turbines to produce electricity. After doing its work, the spent steam must be cooled and condensed back into water. To condense the steam, cool water is brought into the plant and circulated next to the hot steam. In this process, the water used for cooling warms 5 to 10 Celsius degrees (9 to 18 Fahrenheit degrees), after which it may be dumped back into the lake, river, or ocean from which it came. Similarly, factories contribute to thermal pollution when they dump water used to cool their machinery.

The second type of thermal pollution is much more widespread. Streams and small lakes are naturally kept cool by trees and other tall plants that block sunlight. People often remove this shading vegetation in order to harvest the wood in the trees, to make room for crops, or to construct buildings, roads, and other structures. Left unshaded, the water warms by as much as 10 Celsius degrees (18 Fahrenheit degrees). In a similar manner, grazing sheep and cattle can strip streamsides of low vegetation, including young trees. Even the removal of vegetation far away from a stream or lake can contribute to thermal pollution by speeding up the erosion of soil into the water, making it muddy. Muddy water absorbs more energy from the sun than clear water does, resulting in further heating. Finally, water running off of artificial surfaces, such as streets, parking lots, and roofs, is warmer than water running off vegetated land and, thus, contributes to thermal pollution.

Impact

All plant and animal species that live in water are adapted to temperatures within a certain range. When water in an area warms more than they can tolerate, species that cannot move, such as rooted plants and shellfish, will die. Species that can move, such as fish, will leave the area in search of cooler conditions, and they will die if they can not find them. Typically, other species, often less desirable, will move into the area to fill the vacancy. In general, cold waters are better habitat for plants and animals than warm ones because cold waters contain more dissolved oxygen. Many freshwater fish species that are valued for sport and food, especially trout and salmon, do poorly in warm water. Some organisms do thrive in warm water, often with undesirable effects. Algae and other plants grow more rapidly in warm water than in cold, but they also die more rapidly; the bacteria that decompose their dead tissue use up oxygen, further reducing the amount available for animals. The dead and decaying algae make the water look, taste, and smell unpleasant.

Effects of a thermal pollution

Heat and hot water result from many industrial processes. They are in particular by-products of the activity of the power stations and nuclear. The water rejected into the marine mediums has harmful effects, primarily on the marine animal-life.The pollution thermal has effects difficult to define with precision. One observes significant interferences of the reduction in the salinity of water and variations of this one in time and space. Several actions of such a pollution were however determined. It was in particular established that the hotter the temperature of rejected water is and more harmful is its effect on the benthic organizations.On the algae, the action of a thermal pollution is very variable. One noted opposite effects is strong proliferations or on the contrary a significant mortality. The metabolism of the algae, the phytoplanktons, the benthic macrophages is very faded by heated water.One will thus be satisfied to point out some facts of observation which seem to have characters of general data:

Ecological effects — warm water

Warm water typically decreases the level of dissolved oxygen in the water. The decrease in levels of dissolved oxygen can harm aquatic animals such as fish, amphibians and copepods. Thermal pollution may also increase the metabolic rate of aquatic animals, as enzyme activity, resulting in these organisms consuming more food in a shorter time than if their environment were not changed. An increased metabolic rate may result in food source shortages, causing a sharp decrease in a population. Changes in the environment may also result in a migration of organisms to another, more suitable environment, and to in-migration of organisms that normally only live in warmer waters elsewhere. This leads to competition for fewer resources; the more adapted organisms moving in may have an advantage over organisms that are not used to the warmer temperatures. As a result one has the problem of compromising food chains of the old and new environments. Biodiversity can be decreased as a result.It is known that temperature changes of even one to two degrees Celsius can cause significant changes in organism metabolism and other adverse cellular biology effects. Principal adverse changes can include rendering cell walls less permeable to necessary osmosis, coagulation of cell proteins, and alteration of enzyme metabolism. These cellular level effects can adversely affect mortality and reproduction.

Primary producers are affected by warm water because higher water temperature increases plant growth rates, resulting in a shorter lifespan and species overpopulation. This can cause an algae bloom which reduces the oxygen levels in the water. The higher plant density leads to an increased plant respiration rate because the reduced light intensity decreases photosynthesis. This is similar to the eutrophication that occurs when watercourses are polluted with leached agricultural inorganic fertilizers.

A large increase in temperature can lead to the denaturing of life-supporting enzymes by breaking down hydrogen- and disulphide bonds within the quaternary structure of the enzymes. Decreased enzyme activity in aquatic organisms can cause problems such as the inability to break down lipids, which leads to malnutrition.In limited cases, warm water has little deleterious effect and may even lead to improved function of the receiving aquatic ecosystem. This phenomenon is seen especially in seasonal waters and is known as thermal enrichment. An extreme case is derived from the aggregational habits of the manatee, which often uses power plant discharge sites during winter. Projections suggest that manatee populations would decline upon the removal of these discharges.

The added heat lowers the dissolved oxygen content and may cause serious problems for the plants and animals living there. In extreme cases, major fish kills can result. Warm water may also increase the metabolic rate of aquatic animals, as enzyme activity, meaning that these organisms will consume more food in a shorter time than if their environment was not changed.

Ecological Effects — Cold Water

Releases of unnaturally cold water from reservoirs can dramatically change the fish and macroinvertebrate fauna of rivers, and reduce river productivity. In Australia, where many rivers have warmer temperature regimes, native fish species have been eliminated, and macroinvertebrate faunas drastically altered and impoverished, from large stretches of rivers due to cold water thermal pollution.

A solution for thermal pollution

Using nearby water for cooling in industrial applications and then returning that water to the source is a traditional maneuver, but it can also damage local aquatic life and violate EPA regulations. Now more than ever, cooling towers are being used to precool that water before discharge, diverting rejected heat into the atmosphere instead. Learn the costs and benefits of a few configurations, along with tips for dealing with blowdown and effluent within the facility itself.Warm water can be found in a number of different areas in nature-hot springs or water warmed by volcanic activity. Warm water can become a problem, however, when it is created by man and introduced into nature. An example of this is when the water used to cool power plants or other industrial applications is discharged into streams, rivers, and lakes.

This is known as thermal pollution or thermal discharge, and it is the introduction of waste heat into bodies of water that support aquatic life. The addition of heat reduces the water's ability to hold dissolved gases, including the oxygen required for aquatic life. If the water temperature is greater than 95[degrees]F, the dissolved oxygen content may be too low to support some species. If the differential temperature is too large, the difference can also stress some species.As a result, thermal pollution can wreak havoc on native fish species, such as trout, that require cold water with high levels of dissolved oxygen. When the water becomes warmer, other non-native fish that thrive in the warmth can take over habitats from native fish. In addition, warmer water allows bacterial populations to increase and thrive, and algae "blooms" may occur.Regulators and lawmakers in the United States long ago recognized that thermal pollution is a problem and addressed the issue of the EPA Clean Water Act. States and other regulatory agencies use those guidelines to require power plants and industries to limit warm water discharges back into surface waters, sometimes by way of cooling towers.

GROWING NEED FOR TOWERS

Geological Survey about 48% of all freshwater and saline-water withdrawals for 2000 were used for thermoelectric power. Most of this water was derived from surface water and used for once through cooling at power plants. About 52% of fresh surface water withdrawals and about 96% of saline water withdrawals were for thermoelectric power use.This large amount of water is needed by power plants due to the fact that over the years, there has been an ever-increasing need for electricity. This means power plants are expected to run at near maximum output for a large part of the year. The cheapest and easiest method for power plants to operate has always been to withdraw water from a nearby body of water, pass it through the plant, and return the heated water to the same body of water.

These once-through cooling systems now require very strict environmental permits, issued in accordance with the National Pollution Discharge Elimination System. The permits vary from state to state and location to location and may have different requirements.Some permits require that the plant must discharge the water within a temperature differential limit over the temperature of the intake water. Other permits have an ultimate limit; in other words, they can't ever exceed a specific temperature. Still other permits have both differential and ultimate limits.Depending on the permit, restrictions are often magnified during low river or lake levels, or drought conditions. That's because most utilities see their peak loads in the summer months, when air conditioning loads are high. In addition, water temperatures are at their highest in the summer, which can make it difficult for power plants to comply with permit requirements.

RADIOACTIVE POLLUTION

Radioactive pollution can be defined as the release of radioactive substances or high-energy particles into the air, water, or earth as a result of human activity, either by accident or by design. The sources of such waste include:

- Nuclear weapon testing or detonation;
- The *nuclear fuel cycle*, including the mining, separation, and production of nuclear materials for use in nuclear power plants or nuclear bombs;
- Accidental release of radioactive material from nuclear power plants. Sometimes natural sources of radioactivity, such as radon gas emitted from beneath the ground, are considered pollutants when they become a threat to human health.

Since even a small amount of radiation exposure can have serious (and cumulative) biological consequences, and since many radioactive wastes remain toxic for centuries, radioactive pollution is a serious environmental concern even though natural sources of radioactivity far exceed artificial ones at present.The problem of radioactive pollution is compounded by the difficulty in assessing its effects. Radioactive waste may spread over a broad area quite rapidly and irregularly from an abandoned dump into an aquifer and may not fully show its effects upon humans and organisms for decades in the form of cancer or other chronic diseases.

Surface waters are a powerful factor that causes migration of radionuclides across the territory of Belarus. For this reason it is essential to take into due account the transit role of rivers in the transportation of radionuclides, including transboundary transfer. In watercourses and flowing water bodies concentration of radionuclides are reducing every year, but they tend to accumulate in sttatic water bodies (lakes, ponds, reservoirs, especially in bottom sediments). Due to the accident at the Chernobyl nuclear power station (CNPS) one quarter of the country's territory was contaminated by caesium-137 (23%), strontium-90 (10%) and plutonium (about 2 %). Concentrations of caesium-137 early 90's presented. According to the monitoring data the radiation situation in the Dnieper-Sozh and Pripyat basins is stable. The annual average concentration of caesium-137 has decreased significantly in large and small rivers for the observed early 1990's period. Exceeding of the National permissible levels for caesium-137 and strontium-90 were not observed.

Caesium-137 concentration in surface waters are closely connected with the annual volume of river flow, as can be seen from the increase in concentrations of caesium-137 in some rivers, where water supply was lower than the perennial average values. The data analysis of concentration of caesium-137 during the spring flood in the Pripyat basin shows that concentration of caesium-137 in a dissolved form in the Pripyat basin remains at the level of the average indices for the prior period. Yet concentration of this radionuclide has considerably increased in dredges. This means that caesium-137 is washed out and transported by flood flow with sediments.

Since the radiation situation has stabilized, transboundary transport of radioactive elements through river flow has significantly decreased. Mainly it is the Pripyat that transports radionuclides, in particular strontium-90, as they are washed out from the 30-kilometer Chernobyl zone. Due attention is devoted to studies of the radiation state of small rivers, which are tributaries of the Pripyat and the Sozh in the most contaminated areas of the Gomel and Mogilev regions. Over the years radioactivity in water tends to decrease. The exception is the dissolved strontium -90, which is a specific feature for the CNPS zone.

Annual data on the content of caesium-137 and strontium-90 (in soluble and suspended forms) in the bottom sediments and water biota suggest that bottom sediments and water biota are significant contributors to the total radioactivity of surface water systems. A tendency towards a reduction in radioactivity of bottom sediments and water biota is minor.

During spring high water and summer-autumn flood, migration of radionuclides into open water systems occurs both in soluble and in absorbed forms on organic and mineral carrieres. The ratio between concentrations of caesium-137 and strontium-90 at the end checkpoints of the Braginka and Senna Rivers suggests that starting from1992-1993, the concentration of strontium-90 has begun to exceed that of caesium-137. That phenomenon is characteristic for surface watercourses close to the CNPS zone and is explained by an increase in migration ability of strontium-90 due to its release from active particles.

RADIOACTIVE MATERIALS REGULATIONS

Many facilities use radioactive material (RAM) in diverse ways or have radioactive wastes.Exit signs may contain tritium or radioactive hydrogen-3 gas, and smoke detectors may contain radium. Other examples of industrial uses of RAM include devices to measure the density of concrete or blacktop, determine the thickness of paper and rolled steel as it is made, find cracks in pipes or airplane surfaces, test the amount of lead in paint, or monitor the flow of sludge through pipes at a sewage treatment plant.

Research facilities and academic institutions use RAM during the development of new pharmaceuticals to "tag" certain molecules to follow their progress through chemical or biological processes and in other research activities. Medical facilities inject patients with RAM to diagnose medical conditions and for therapeutic treatments. Medical facilities also use large radiation sources for cancer treatment.

Radium paint was once used on aircraft instruments, naval compasses, military vehicle instruments, and on clocks and watches to make the numbers and lines glow in the dark. Naturally occurring radioactive material is found as uranium in clay and bricks, granite, shale, or other rocks. It is also found as radium in soils or as radium sulfate scales on some pipes and fittings from the oil and gas industry and as the naturally radioactive constituent of potassium, potassium-40.

Environmental Monitoring and Radon Gas

The Environmental Monitoring Unit (EMU) of the RPMWS is responsible for DEQ's Radon Progamme activities. Environmental Protection Agency's State Indoor Radon Grant Progamme. This progamme provides education on the radiological risks posed by public exposure to radon gas and works closely with local health departments throughout the state for outreach at the local level.The EMU operates an environmental monitoring network around each of Michigan's nuclear power plant sites.

The progamme collects and performs radioanalyses of several types of samples, including direct radiation, air, surface water, precipitation, and milk from the environs of the nuclear plants. Unit labouratory analyses also include samples collected by other section staff during investigations of potentially contaminated sites, during emergency response activities, and from routine staff compliance investigations.

Nuclear Facilities

The Nuclear Facilities Unit of the RPMWS develops and implements the DEQ's Nuclear Facilities Emergency Response Procedures and the nuclear accident aspects of the Michigan Emergency Management Plan as they relate to the DEQ's responsibilities to respond to accidents or emergencies at any of Michigan's commercial nuclear power plants or to large-scale radiological incidents. These efforts are conducted in cooperation with other state agencies and under the overall emergency response coordination of the Michigan State Police.

Unit staff also interact with nuclear plant utility staff and staff of the NRC concerning the day-to-day operations of nuclear power reactors to assure radiological protection of the public and the environment.

RADIOACTIVE MATERIALS REGULATIONS

Waste Industrial Smoke Detectors

Remove any batteries from the detector and handle the battery as a universal waste or under the applicable hazardous waste regulations for that company's hazardous waste generator status

The specific requirements a company would have to follow would depend on if the smoke detectors were subject to the federal nuclear regulations or if it was a hazardous waste. There are two types of materials commonly found in used smoke detectors.

Radioactive Materials Regulations

- The older models may contain non-exempt radium-226 sources that are regulated under the state of Michigan. These detectors should not go to a solid waste landfill but returned to the manufacturer or disposed as radioactive waste.
- The newer models contain a small americium source. The combined smoke detector and americium source have a specific exemption in the federal regulations. Large quantities such as resulting from a major construction renovation project should not be disposed without first checking with officials of the NRC or the WHMD radioactive materials staff.

If the smoke detectors are not recycled for metal, some smoke detectors could be subject to the hazardous waste regulations because the amount of metal in the detectors may fail the Toxicity Characteristic Leaching Procedure. Small quantity generators and large quantity generators can not put hazardous waste smoke detectors in the trash. Conditionally exempt small quantity generators may dispose smoke detectors in licensed solid waste landfills if the landfill will accept them. If these smoke detectors are not classified as a hazardous waste, then they may be sent to a licensed landfill. However, companies should contact the landfill if disposing of large numbers because the waste load may set off the landfill's radiation detectors.

Contact the DEQ Waste and Hazardous Materials Division regarding potential safety concerns when numerous smoke detectors are disposed of at the same time or regarding nuclear regulations.

The inspection and mapping of radioactive sources

During the ten years since the Norwegian-Russian cooperation was established in form of The Group of Experts, a magnificent effort in mapping the present environmental situation has been carried out. That includes sources of radioactivity in the environment

On these missions water samples were collected, along with sediments and living organisms which have given new knowledge about the present environmental status of the area. At two occasions closer investigations were done at dumping places for radioactive materials, and due to this work future risk of pollution has been evaluated.

At the moment effort is being made to establish a co-ordinated progamme for surveying the northern sea areas, a system for exchanging data, and an extension of a network for automatic metering in the county of Murmansk. Experiences from the cooperation in the Expert Group gave a solid foundation for joint inspections and risk evaluation

In addition to the work done by the Expert group, inspections of different kinds of pollution in the Arctic are carried out by the organizations called AMAP (Arctic Monitoring and Assessment Progammeme). AMAP has established an international data centre, localized at the Norwegian Radiation Protection Authority (NRPA) for the purpose of monitoring sources and levels of radioactive pollution. The work of AMAP is close to the subjects investigated by the Expert group, resulting in close contact between the two organisations.

RISK EVALUATION

The joint projects between Norway and Russia include risk analysis of potential scenes of accidents at several nuclear plants in Russia. Substantial amounts of radioactive waste was released from nuclear plants and directed out in rivers and ocean, causing severe pollution of the environment. Today, storing of radioactive waste at plants with poor securing is a big problem. There is great concern about the consequences if accidents take place, what the effects might be on the local environment, and even for other countries.

Calculations and risk evaluations have been made in case of potential accidents occur at the nuclear plant in Majak in Ural. Majak is most likely the most contaminated area caused by previous unfortunate incidents and releases to the environment. In management of the Expert group, fieldwork has been carried out twice, and potential accidents have been considered. Two of the most important potential scenarios at Majak are an explosion in a container for high-level waste, and destruction of the lower dam at the plant. This scenario has potential for large releases of radioactive substances with scattering by the river Ob to the Kara Sea, but the level of pollution will be relatively limited. A similar evaluation is being done at the nuclear plant of Krasnojarsk close to the river Jenisej. Norwegian authorities have given special attention to the nuclear plant at Kola. This plant is placed 200 km from the Norwegian border. Analysis has been made on long-term effects on the environment and population in the Nort-West of Russia from possible accidents at this plant. The worst thinkable scenarios have been used in the modelling; meaning large fallouts, and bad weather conditions during an accident. The results indicate that the negative effects will be long lasting.

Radioactive Mine Waste Polluting Colourado River:

Water tests reveal that "uranium mill waste" leaching into the Colourado River has made the water radioactive at "one-third the level considered dangerous," says the San Diego Union Tribune 1/10. The mine's owner, Atlas Corporation, has declared bankruptcy, leaving the bulk of the enormous clean up costs to taxpayers. The huge pile of mine waste "sits 750 feet from the river," and is leaking "an estimated 28,800 gallons of radioactive pollution and toxic chemicals" into the river each day.

TYPES OF RADIATION

Radiation is classified as being ionizing or nonionizing. Both types can be harmful to humans and other organisms.

Radioactive Pollution - Nonionizing Radiation

Radioactive Pollution - Ionizing Radiation

Radioactive Pollution - Sources of Radioactive Pollution

Radioactive Pollution - Lifestyle and Radiation Dose

Radioactive Pollution - Nuclear Weapons Testing

Radioactive Pollution - Nuclear Power Plants

Radioactive Pollution - Biological Effects of Radioactivity

Nonionizing radiation

Nonionizing radiation is relatively long-wavelength electromagnetic radiation, such as radio waves, microwaves, visible radiation, ultraviolet radiation, and very low-energy electromagnetic fields. Nonionizing radiation is generally considered less dangerous than ionizing radiation. However, some forms of nonionizing radiation, such as ultraviolet, can damage biological molecules and cause health problems. Scientists do not yet fully understand the longer-term health effects of some forms of nonionizing radiation, such as that from very low-level electromagnetic fields (e.g., high-voltage power lines), although the evidence to date suggests that the risks are extremely small.

Ionizing radiation

Ionizing radiation is the short wavelength radiation or particulate radiation emitted by certain unstable isotopes during radioactive decay. There are about 70 radioactive isotopes, all of which emit some form of ionizing radiation as they decay from one isotope to another. A radioactive isotope typically decays through a series of other isotopes until it reaches a stable one. As indicated by its name, ionizing radiation can ionize the atoms or molecules with which it interacts. In other words, ionizing radiation can cause other atoms to release their electrons. These free electrons can damage many biochemicals, such as proteins, lipids, and nucleic acids (including DNA). In intense, this damage can cause severe human health problems, including cancers, and even death.

Ionizing radiation can be either short-wavelength electromagnetic radiation or particulate radiation. Gamma radiation and X-radiation are short-wavelength electromagnetic radiation. Alpha particles, beta particles, neutrons, and protons are particulate radiation. Alpha particles, beta particles, and gamma rays are the most commonly encountered forms of radioactive pollution. Alpha particles are simply ionized helium nuclei, and consist of two protons and two neutrons. Beta particles are electrons, which have a negative charge. Gamma radiation is high-energy electromagnetic radiation.

Scientists have devised various units for measuring radioactivity. A Curie (Ci) represents the rate of radioactive decay. One Curie is 3.7×10^{10} radioactive disintegrations per second. A rad is a unit representing the absorbed dose of radioactivity. One rad is equal to an absorbed energy dose of 100 ergs per gram of radiated medium. One rad = 0.01 Grays. A rem is a unit that measures the effectiveness of radioactivity in causing biological damage. One rem is equal to one rad times a biological weighting factor. The weighting factor is 1.0 for gamma radiation and beta particles, and it is 20 for alpha particles. One rem = 1000 millirem = 0.01 Sieverts. The radioactive half-life is a measure of the persistence of radioactive material. The half-life is the time required for one-half of an initial quantity of atoms of a radioactive isotope to decay to a different isotope.

Sources of radioactive pollution

In the United States, people are typically exposed to about 350 millirems of ionizing radiation per year. On average, 82% of this radiation comes from natural sources and 18% from anthropogenic sources (i.e., those associated with human activities). The major natural source of radiation is radon gas, which accounts for about 55% of the total radiation dose. The principal anthropogenic sources of radioactivity are medical X-rays and nuclear medicine. Radioactivity from the fallout of nuclear weapons testing and from nuclear power plants make up less than 0.5% of the total radiation dose, i.e., less than 2 millirems. Although the contribution to the total human radiation dose is extremely small, radioactive isotopes released during previous atmospheric testing of nuclear weapons will remain in the atmosphere for the next 100 years.

Lifestyle and radiation dose

People who live in certain regions are exposed to higher doses of radiation.Residents of the Rocky Mountains of Colourado receive about 30 millirems more cosmic radiation than people living at sea level. This is because the atmosphere is thinner at higher elevations, and therefore less effective at shielding the surface from cosmic radiation. Exposure to cosmic radiation is also high while people are flying in an airplane, so pilots and flight attendants have an enhanced, occupational exposure. In addition, residents of certain regions receive higher doses of radiation from radon-222, due to local geological anomalies. Radon-222 is a colourless and odorless gas that results from the decay of naturally occurring, radioactive isotopes of uranium. Radon-222 typically enters buildings from their basement, or from certain mineral-containing construction materials. Ironically, the trend toward improved home insulation has increased the amount of radon-222 which remains trapped inside houses.

Personal lifestyle also influences the amount of radioactivity to which people are exposed.Miners, who spend a lot of time underground, are exposed to relatively high doses of radon-222 and consequently have relatively high rates of lung cancer. Cigarette smokers expose their lungs to high levels of radiation, since tobacco plants contain trace quantities of polonium-210, lead-210, and radon-222. These radioactive isotopes come from the small amount of uranium present in fertilizers used to promote tobacco growth. Consequently, the lungs of a cigarette smoker are exposed to thousands of additional millirems of radioactivity, although any associated hazards are much less than those of tar and nicotine.

Nuclear weapons testing

Nuclear weapons release enormous amounts of radioactive materials when they are exploded. Most of the radioactive pollution from nuclear weapons testing is from iodine-131, cesium-137, and strontium-90. Iodine-131 is the least dangerous of these isotopes, although it has a relatively half-life of about eight days. Iodine-131 accumulates in the thyroid gland, and large doses can cause thyroid cancer. Cesium-137 has a half-life of about 30 years. It is chemically similar to potassium, and is distributed throughout the human body. Based on the total amount of cesium already in the atmosphere, all humans will receive about 27 millirems of radiation from cesium-137 over their lifetime. Strontium-90 has a half-life of 38 years. It is chemically similar to calcium and is deposited in bones. Strontium-90 is expelled from the body very slowly, and the uptake of significant amounts increases the risks of developing bone cancer or leukemia.

Nuclear power plants

Many environmentalists are critical of nuclear power generation. They claim that there is an unacceptable risk of a catastrophic accident, and that nuclear power plants generate large amounts of unmanageable nuclear waste.

Nuclear Regulatory Commission has strict requirements regarding the amount of radioactivity that can be released from a nuclear power reactor. In particular, a nuclear reactor can expose an individual who lives on the fence line of the power plant to no more than 10 millirems of radiation per year. Actual measurements at U.S. nuclear power plants have shown that a person who lived at the fence line would actually be exposed to much less that 10 millirems.Thus, for a typical person who is exposed to about 350 millirems of radiation per year from all other sources, much of which is natural background, the proportion of radiation from nuclear power plants is extremely small. In fact, coal- and oil-fired power plants, which release small amounts of radioactivity contained in their fuels.

Although a nuclear power plant cannot explode like an atomic bomb, accidents can result in serious radioactive pollution. During the past 45 years, there have been a number of not-fully controlled or uncontrolled fission reactions at nuclear power plants in the United States and elsewhere, which have killed or injured power plant workers. These accidents occurred in Los Alamos, New Mexico; Oak Ridge, Tennessee; Richland, Washington; and Wood River Junction, Rhode Island. The most famous case was the 1979 accident at the Three Mile Island nuclear reactor in Pennsylvania, which received a great deal of attention in the press.

However, nuclear scientists have estimated that people living within 50 mi (80 km) of this reactor were exposed to less than two millirems of radiation, most of it as iodine-131, a short-lived isotope. This exposure constituted less than 1% of the total annual radiation dose of an average person. However, these data do not mean that the accident at Three Mile Island was not a serious one; fortunately, technicians were able to reattain control of the reactor before more devastating damage occurred, and the reactor system was well contained so that only a relatively small amount of radioactivity escaped to the ambient environment.

By far, the worst nuclear reactor accident occurred in 1986 in Chernobyl, Ukraine. An uncontrolled build-up of heat resulted in a meltdown of the reactor core and combustion of graphite moderator material in one of the several generating units at Chernobyl, releasing more than 50 million Curies of radioactivity to the ambient environment. The disaster killed 31 workers, and resulted in the hospitalization of more than 500 other people from radiation sickness. According to Ukrainian authorities, during the decade following the Chernobyl disaster an estimated 10,000 people in Belarus, Russia, and Ukraine died from cancers and other radiation-related diseases caused by the accident. In addition to these relatively local effects, the atmosphere transported radiation from Chernobyl into Europe and throughout the Northern Hemisphere.

More than 500,000 people in the vicinity of Chernobyl were exposed to dangerously high doses of radiation, and more than 300,000 people were permanently evacuated from the vicinity. Since radiation-related health problems may appear decades after exposure, scientists expect that many thousands of additional people will eventually suffer higher rates of thyroid cancer, bone cancer, leukemia, and other radiation-related diseases. Unfortunately, a cover-up of the explosion by responsible authorities, including those in government, endangered even more people. Many local residents did not known that they should flee the area as soon as possible, or were not provided with the medical attention they needed.

The large amount of radioactive waste generated by nuclear power plants is another an important problem. This waste will remain radioactive for many thousands of years, so technologists must design systems for extremely long-term storage. One obvious problem is that the long-term reliability of the storage systems cannot be fully assured, because they cannot be directly tested for the length of time they will be used (i.e., for thousands of years). Another problem with nuclear waste is that it will remain extremely dangerous for much longer than the expected lifetimes of existing governments and social institutions. Thus, we are making the societies of the following millennia, however they may be structured, responsible for the safe storage of nuclear waste that is being generated in such large quantities today.

Biological effects of radioactivity

The amount of injury caused by a radioactive isotope depends on its physical half-life, and on how quickly it is absorbed and then excreted by an organism. Most studies of the harmful effects of radiation have been performed on single-celled organisms. Obviously, the situation is more complex in humans and other multicellular organisms, because a single cell damaged by radiation may indirectly affect other cells in the individual. The most sensitive regions of the human body appear to be those which have many actively dividing cells, such as the skin, gonads, intestine, and tissues that grow blood cells (spleen, bone marrow, lymph organs).

Radioactivity is toxic because it forms ions when it reacts with biological molecules. These ions can form free radicals, which damage proteins, membranes, and nucleic acids. Radioactivity can damage DNA (deoxyribonucleic acid) by destroying individual bases (particularly thymine), by breaking single strands, by breaking double strands, by cross-linking different DNA strands, and by cross-linking DNA and proteins. Damage to DNA can lead to cancers, birth defects, and even death.

However, cells have biochemical repair systems which can reverse some of the damaging biological effects of low-level exposures to radioactivity. This allows the body to better tolerate radiation that is delivered at a low dose rate, such as over a longer period of time. In fact, all humans are exposed to radiation in extremely small doses throughout their life. The biological effects of such small doses over such a long time are almost impossible to measure, and are essentially unknown at present. There is, however, a theoretical possibility that the small amount of radioactivity released into the environment by normally operating nuclear power plants, and by previous atmospheric testing of nuclear weapons, has slightly increased the incidence of certain cancers in human populations. However, scientists have not been able to conclusively show that such an effect has actually occurred.

Currently, there is disagreement among scientists about whether there is a threshold dose for radiation damage to organisms. In other words, is there a dose of radiation below which there are no harmful biological effects? Some scientists maintain that there is no such threshold, and that radiation at any dose carries a finite risk of causing some biological damage. Furthermore, the damage caused by very low doses of radiation may be cumulative, or additive to the damage caused by other harmful agents to which humans are exposed.

Other scientists maintain that there is a threshold dose for radiation damage. They believe that biological repair systems, which are presumably present in all cells, can fix the biological damage caused by extremely low doses of radiation. Thus, these scientists claim that the extremely low doses of radiation to which humans are commonly exposed are not harmful.

One of the most informative studies of the harmful effects of radiation is a long-term investigation of the survivors of the 1945 atomic blasts at Hiroshima and Nagasaki by James Neel and his colleagues. The survivors of these explosions had abnormally high rates of cancer, leukemia, and other diseases. However, there seemed to be no detectable effect on the occurrence of genetic defects in children of the survivors. The radiation dose needed to cause heritable defects in humans is higher than biologists originally expected.

Radioactive pollution is an important environmental problem. It could become much worse if extreme vigilance is not utilized in the handling and use of radioactive materials, and in the design and operation of nuclear power plants.

RADIATION EXPOSURE

Radiation exposure occurs any time that energy in the form of electromagnetic rays or particles interacts with biological tissue. Ionizing radiation is particularly energetic; examples include: x rays, gamma radiation, and subatomic particles. Biological damages caused by exposure to ionizing range from mild tissue burns to cancer, genetic damage, and ultimately, death. However, there are potential benefits of controlled exposures to certain kinds of radiation, which can be used for the detection, diagnosis, and treatment of certain diseases. Exposure to many types of radiation is routinely monitored using sensitive devices, such as film badges and dosimeters.

Effects of radiation exposure

With many forms of ionizing radiation, energy is transferred to electrons that surround atomic nuclei. Atoms affected by x rays usually absorb enough energy to lose some of their electrons, and so become ionized. (An atom is ionized when it gains or loses electrons and acquires a net electric charge.) Ultraviolet radiation causes electrons to absorb energy and jump to a higher energy orbit around the atomic nucleus. The sun and sunlamps emit enough ultraviolet radiation to cause sunburn, premature aging of the skin, and skin cancers. Exposure of humans and animals to ultraviolet radiation also results in the production of vitamin D, a biochemical necessary for good health.

Radiation that consists of charged particles can knock electrons out of their orbit around atoms. This also creates ions. Such radiation can also cause atoms to enter an exited state, if the electrons are bumped into higher-energy orbits. These changes result in atoms and molecules (including biochemicals) that are chemically reactive. Seeking to become stable, they interact with unaffected atoms and molecules, which may be damaged (i.e., changed) in the process, and are then unable to perform their usual metabolic functions. Nuclear material such as DNA molecules may be damaged to the degree that they can no longer be accurately copied. This may lead to impaired cell function, cell death, or genetic abnormalities.

RADIOACTIVE FALLOUT

Radioactive fallout is material produced by a nuclear explosion or a nuclear reactor accident that enters the atmosphere and eventually falls to Earth. This fallout consists of minute, radioactive particles of dust, soil, and other debris. While some fallout results from natural sources, the term is usually used in reference to radioactive particles that were released into the atmosphere by a nuclear explosion or reactor accident. Fallout refers to material that has fallen to Earth, and also to material that is still suspended in the atmosphere.

Types of fallout

Particles that make up radioactive fallout can be as small as the invisible droplets produced by an aerosol spray can, or as large as ash that falls close to a wood fire. The type of radioactivity in fallout depends on the nature of the nuclear reaction that emitted the particles into the atmosphere. More than 60 different types of radioactive substances may be initially present in fallout. Some of these decay into non-radioactive products in seconds, while others take centuries or longer to become non-radioactive. It takes 28 years, for a sample of strontium-90 to lose one-half of its initial radioactivity. Strontium-90 is one of the most dangerous elements in fallout because it is treated by the metabolism of humans in the same manner as calcium, an important component of bone. If animals or humans eat food contaminated with strontium-90, it will accumulate in their bodies. Other particularly harmful products in fallout include cesium-134, cesium-137, and iodine-131.Radiation damages and kills cells in the body. Large doses of radiation can result in burns, vomiting, and damage to the nervous system, digestive system, and bone marrow. Smaller doses can cause genetic mutations and cancer years after exposure.

2

Biodegradation of Waste

BIODEGRADATION OF WASTE

Biodegradation is nature's way of recycling wastes, or breaking down organic matter into nutrients that can be used by other organisms. "Degradation" means decay, and the "bio-" prefix means that the decay is carried out by a huge assortment of bacteria, fungi, insects, worms, and other organisms that eat dead material and recycle it into new forms.

In nature, there is no waste because everything gets recycled. The waste products from one organism become the food for others, providing nutrients and energy while breaking down the waste organic matter. Some organic materials will break down much faster than others, but all will eventually decay.

By harnessing these natural forces of biodegradation, people can castes and clean up some types of environmental contaminants. Through composting, we accelerate natural biodegradation and convert organic wastes to a valuable resource. Wastewater treatment also accelerates natural forces of biodegradation. In this case the purpose is to break down organic matter so that it will not cause pollution problems when the water is released into the environment. Through bioremediation, microorganisms are used to clean up oil spills and other types of organic pollution. Composting and bioremediation provide many possibililites for student research.

Wastewater Treatment

Wastewater treatment uses microbes to decompose organic matter in sewage. If too much untreated sewage or other organic matter is added to a lake or stream, dissolved oxygen levels will drop too low to support sensitive species of fish and other aquatic life. Wastewater treatment systems are designed to digest much of the organic matter before the wastewater is released so that this will not occur. Treatment systems use physical, chemical, and biological processes:

- Primary treatment physically removes large solids using grates, screens, and settling tanks.
- Secondary treatment promotes growth of bacteria and other microbes that break down the organic wastes. These biodegradation processes also take place in streams, lakes, and oceans, but the purification systems in nature can easily be overloaded with input of too much organic waste. Secondary treatment prevents this type of pollution by degrading most of the organic matter before the water is released into the environment.
- Tertiary treatment is used only where it is needed to protect the receiving waters from excess nutrients. In tertiary treatment, the concentrations of phosphorus or nitrogen are reduced through biological or chemical processes.
- Disinfection kills disease-causing organisms, most commonly through chlorination.

Sludge, the collection of solids that are removed during wastewater treatment, requires processing to reduce odor and water content. Depending on the disposal method, the sludge also may undergo treatment to decompose organic matter or kill disease-causing organisms.

Wastewater Treatment Principles

Sewage is the wastewater released by residences, businesses and industries in a community. It is 99.94 percent water, with only 0.06 percent of the wastewater dissolved and suspended solid material. The cloudiness of sewage is caused by suspended particles which in untreated sewage ranges from 100 to 350 mg/l. A measure of the strength of the wastewater is biochemical oxygen demand, or BOD_5. The BOD_5 measures the amount of oxygen microorganisms require in five days to break down sewage. Untreated sewage has a BOD_5 ranging from 100 mg/l to 300 mg/l. Pathogens or disease-causing organisms are present in sewage. Coliform bacteria are used as an indicator of disease-causing organisms. Sewage also contains nutrients (such as ammonia and phosphorus), minerals, and metals. Ammonia can range from 12 to 50 mg/l and phosphorus can range from 6 to 20 mg/l in untreated sewage.

Sewage treatment is a multi-stage process to renovate wastewater before it reenters a body of water, is applied to the land or is reused. The goal is to reduce or remove organic matter, solids, nutrients, disease-causing organisms and other pollutants from wastewater. Each receiving body of water has limits to the amount of pollutants it can receive without degradation. Therefore, each sewage treatment plant must hold a permit listing the allowable levels of BOD_5, suspended solids, coliform bacteria and other pollutants. The discharge permits are called NPDES permits which stands for the National Pollutant Discharge Elimination System.

Preliminary Treatment

Preliminary treatment to screen out, grind up, or separate debris is the first step in wastewater treatment. Sticks, rags, large food particles, sand, gravel, toys, etc., are removed at this stage to protect the pumping and other equipment in the treatment plant. Treatment equipment such as bar screens, comminutors (a large version of a garbage disposal), and grit chambers are used as the wastewater first enters a treatment plant. The collected debris is usually disposed of in a landfill.

Primary Treatment

Primary treatment is the second step in treatment and separates suspended solids and greases from wastewater. Waste-water is held in a quiet tank for several hours allowing the particles to settle to the bottom and the greases to float to the top. The solids drawn off the bottom and skimmed off the top receive further treatment as sludge. The clarified wastewater flows on to the next stage of wastewater treatment. Clarifiers and septic tanks are usually used to provide primary treatment.

Secondary Treatment

Secondary treatment is a biological treatment process to remove dissolved organic matter from wastewater. Sewage microorganisms are cultivated and added to the wastewater. The microorganisms absorb organic matter from sewage as their food supply. Three approaches are used to accomplish secondary treatment; fixed film, suspended film and lagoon systems.

Fixed Film Systems

Fixed film systems grow microorganisms on substrates such as rocks, sand or plastic. The wastewater is spread over the substrate, allowing the wastewater to flow past the film of microorganisms fixed to the substrate. As organic matter and nutrients are absorbed from the wastewater, the film of microorganisms grows and thickens. Trickling filters, rotating biological contactors, and sand filters are examples of fixed film systems.

Suspended Film Systems

Suspended film systems stir and suspend microorganisms in wastewater. As the microorganisms absorb organic matter and nutrients from the wastewater they grow in size and number. After the microorganisms have been suspended in the wastewater for several hours, they are settled out as a sludge. Some of the sludge is pumped back into the incoming wastewater to provide "seed" microorganisms. The remainder is wasted and sent on to a sludge treatment process. Activated sludge, extended aeration, oxidation ditch, and sequential batch reactor systems are all examples of suspended film systems.

Lagoon Systems

Lagoon systems are shallow basins which hold the wastewater for several months to allow for the natural degradation of sewage. These systems take advantage of natural aeration and microorganisms in the wastewater to renovate sewage.

Final Treatment

Final treatment focuses on removal of disease-causing organisms from wastewater. Treated wastewater can be disinfected by adding chlorine or by using ultraviolet light. High levels of chlorine may be harmful to aquatic life in receiving streams. Treatment systems often add a chlorine-neutralizing chemical to the treated wastewater before stream discharge.

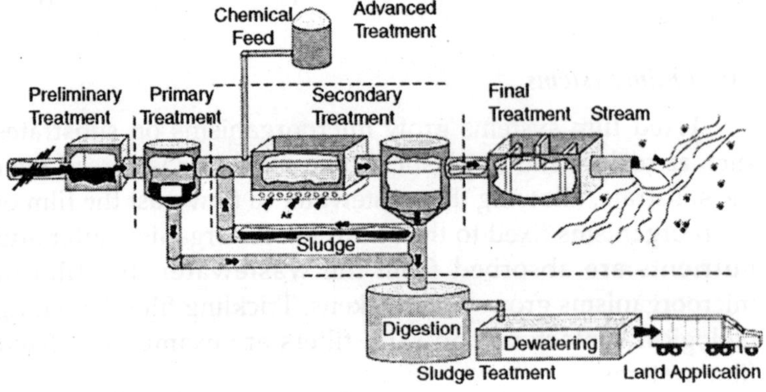

Fig.: Sludge Treatment

Advanced Treatment

Advanced treatment is necessary in some treatment systems to remove nutrients from wastewater. Chemicals are sometimes added during the treatment process to help settle out or strip out phosphorus or nitrogen. Some examples of nutrient removal systems include coagulant addition for phosphorus removal and air stripping for ammonia removal.

Sludges

Sludges are generated through the sewage treatment process. Primary sludges, material that settles out during primary treatment, often have a strong odor and require treatment prior to disposal. Secondary sludges are the extra microorganisms from the biological treatment processes. The goals of sludge treatment are to stabilize the sludge and reduce odors, remove some of the water and reduce volume, decompose some of the organic matter and reduce volume, kill disease causing organisms and disinfect the sludge.

Untreated sludges are about 97 percent water. Settling the sludge and decanting off the separated liquid removes some of the water and reduces the sludge volume. Settling can result in a sludge with about 96 to 92 percent water. More water can be removed from sludge by using sand drying beds, vacuum filters, filter presses, and centrifuges resulting in sludges with between 80 to 50 percent water. This dried sludge is called a sludge cake. Aerobic and anaerobic digestion are used to decompose organic matter to reduce volume. Digestion also stabilizes the sludge to reduce odors. Caustic chemicals can be added to sludge or it may be heat treated to kill disease-causing organisms. Following treatment, liquid and cake sludges are usually spread on fields, returning organic matter and nutrients to the soil.

Wastewater treatment processes require careful management to ensure the protection of the water body that receives the discharge. Trained and certified treatment plant operators measure and monitor the incoming sewage, the treatment process and the final effluent.

Wastewater Treatment Regulations

Clean water has been a concern nationwide Congress adopted the Clean Water Act to protect the waters of the nation. Through this act. Environmental Protection Agency and corresponding state agencies were given the responsibility to regulate activities that threaten the quality of the nation's water resources.

In the Federal Clean Water Act, Congress adopted a comprehensive water policy for the nation and set as a national goal the elimination of pollutant discharges to the navigable waters. An interim goal was set to insure that all navigable waters would be fishable and swimmable

To reach these goals

- No one has the right to pollute the navigable waters of the United States. Dischargers are required to obtain permits.
- Permits shall set limits on the concentration of the pollutants being discharged. A violation of the limits carries a penalty of fines or imprisonment.
- The best technology available shall be used to control the discharge of pollutants.

Each state, through a designated regulatory agency, issues discharge permits and enforces the discharge limits.

The Ohio Water Quality Standards specify that all surface waters of the state must be free from the following pollutants as a result of human activity: suspended solids, floating debris, colour, odor, toxic substances and nutrients that create nuisance growths of weeds and algae.

Permits for wastewater treatment systems serving single family homes, two- or three-family dwellings are issued by local health departments. Permits for all wastewater treatment systems serving more than 2 homes or buildings larger than a 3-family dwelling are issued by the Ohio Environmental Protection Agency.

Two different permits are issued by the Ohio Environmental Protection Agency. All systems must first obtain a Permit to Install (PTI) before a treatment plant can be constructed. The Permit to Install is issued after the plans for construction, operation and management are examined to insure that the receiving stream will be protected.

If the discharge from the treatment plant enters a stream, a NPDES permit is required. The NPDES permit specifies the maximum allowable level of total suspended solids, biochemical oxygen demand, nutrients and bacteria that can be discharged to a stream as well as the minimum level of dissolved oxygen that must be present in the discharge. The levels specified in the NPDES permit are determined by the condition of the receiving stream. Therefore, NPDES permits are subject to change every 5 years as water quality concerns change throughout Ohio.

Discharging raw sewage from a dwelling is considered a public nuisance. For systems serving a 1-, 2- or 3-family dwelling, the local health department is the responsible regulatory agency. The penalty for creating a public nuisance is a third degree misdemeanor which is punishable by not more than 60 days in jail and/or a $500 fine. In addition, the violator may be required to remove the sewage from public or private property or the waters of the state.

Discharges of pollutants from a multiple dwelling is under the jurisdiction of the Ohio Environmental Protection Agency. Discharges may be liable for civil penalties of up to $10,000 for each day of violation. In addition, criminal penalties can be assessed or up to one year of imprisonment or both. A "Connection Ban" can also be issued by the Ohio Environmental Protection Agency which prohibits the construction or installation of home sewage disposal systems when it is shown that pollutants from new homes will be discharged into the waters of the state.

The Ohio Environmental Protection Agency issued 25 administrative orders totaling $80,000 in fines to Ohio communities for wastewater treatment violations.

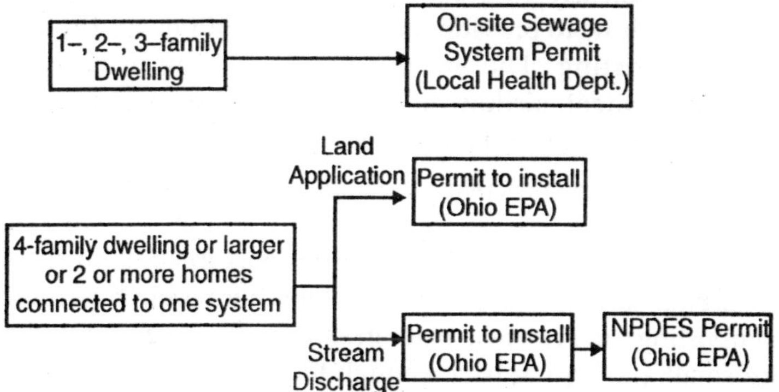

This publication may contain pesticide recommendations that are subject to change at any time. These recommendations are provided only as a guide. It is always the pesticide applicator's responsibility, by law, to read and follow all current label directions for the specific pesticide being used. Due to constantly changing labels and product registrations, some of the recommendations given in this writing may no longer be legal by the time you read them. If any information in these recommendations disagrees with the label, the recommendation must be disregarded.

Composting

Composting refers to the process of combining organic materials such as weeds, grass clippings, and food scraps under conditions that enhance the rate at which they decompose. In the United States, organic wastes make up 20% to 40% of what we throw away. Composting provides an opportunity to keep these organic materials out of landfills and incinerators, using them instead to create a useful end product.

Of course, composting isn't as easy as simply throwing organic waste into big piles and waiting for it to decompose. If moist food scraps or grass clippings are placed in a container and left to sit for a week or two, the result is likely to be a stinky mess that attracts flies. Given the proper conditions, these same organic materials can be composted to produce a material that looks and smells like rich soil and is prized by gardeners. Clearly the second option is preferable, and the key question is how to provide the proper conditions for problem-free composting. The answer lies in paying attention to moisture content, airflow, and the mixture between materials that are high in carbon and those that are high in nitrogen..

Composting takes place at a wide range of scales, from huge commercial or municipal operations to simple backyard bins or heaps. It can also be carried out indoors in containers such as garbage can bioreactors or worm bins. For research purposes, composting can even take place in soda bottle bioreactors, which are small and inexpensive enough to enable students to build their own individualized systems. Once made, compost can be used for gardening projects or for plant growth experiments ranging from nutrient analysis of compost-enriched soils to use of composts to suppress plant diseases

Compost Ingredients

Garden supply stores and catalogs often sell compost "starters," which supposedly speed up the composting process. Develop a recipe for a compost starter and design a research project to test its effect on the compost temperature profile.

How well do human nutrition concepts apply to compost microorganisms? The microbes get a "sugar high," demonstrated by a quick, high temperature peak when fed sugary foods, compared with a longer but lower peak for more complex carbohydrates.

Measure the pH of a number of different compost mixes. How does the pH of initial ingredients affect the pH of finished compost?

Some instructions call for adding lime to increase the pH when compost ingredients are mixed. Other instructions caution to avoid this because it causes a loss of nitrogen. How does adding various amounts of lime to the initial ingredients affect the pH of finished compost.

Microorganisms

Composting recipes sometimes call for inoculating the pile by mixing in a few handfuls of finished compost. Is there any observable difference in appearance of microbes between systems that have and have not been inoculated.

The pH of the initial compost ingredients affect the populations of microorganisms during composting.

Worm Composting

Do organic wastes in compost break down more readily in the presence of worms than through composting that depends solely on microbial decomposition.

In some experiments, plants have not show increased growth when planted in fresh worm castings. Does aging or "curing" worm castings increase their ability to enhance plant growth? Are there chemical differences between fresh and older worm castings? Should worm compost be mixed with soil before being used to grow plants.

Different food sources affect reproductive and growth rates of red worms (*Eisenia fetida*).

Red worms grow best in wastes with pH between 5.0 and 8.0.

Effects of Compost on Plant Growth

Some leaves, such as those of black walnut or eucalyptus trees, contain chemicals that inhibit growth of other plants. Are these compounds broken down by composting.

Finished compost is near neutral pH. Water in which compost has been soaked (often called compost tea) is said to be beneficial to plants. Can you design experiments to test whether different types, concentrations, and amounts of compost tea enhance plant growth?

In China, farmers dig parallel trenches and fill them with organic wastes mixed with cocoons of Eisenia fetida. Soybeans planted in rows between the trenches are highly productive.

Building a Two-Can Bioreactor

Purpose

Two-can bioreactors are designed to be used as small-scall indoor composting units for families, and for composting as an educational tool in the classroom.

Materials

- 32-gallon plastic garbage can
- 20-gallon plastic garbage can
- Drill
- Brick
- Spigot (optional)
- Duct tape (optional)
- Insulation (optional)

Construction

- Using a drill, make 15 to 20 holes (0.5" to 1" diameter) through the bottom of the 20-gallon can. Next drill three rows of holes through the sides of this can, six to eight inches apart with four to five inches between rows, ending about two inches below where the can expands at the top.
- Place a brick or some other object in the bottom of the 32-gallon can. This is to separate the leachate from the compost and allow for its measurement and addition back into the compost pile. The leachate, often referred to as "compost tea," is rich in nutrients which may be in a form readily usable by plants. If not used right away on growing plants, pour the leachate back into the compost. Excessive leachate can be responsible for foul smells. If your sytem produces enough leachage to cause odor problems, your initialscompost mixture was probably too wet.
- Variations on the design:
  - Add insulation to the barrels (inner and outer) with duct tape.
  - Include a spigot to draw off the leachate.
  - Add a layer of old compost, wood chips, or soil inside the outer barrel. This will allow the leachate to be absorbed and may cause fewer leachate/odor issues.

A system of 10-gallon plastic garbage cans that can fit inside 20-gallon cans can be substituted if space is a problem. The smaller system may operate at lower temperatures. This should not affect the final product; it will just take longer before the product can be used.

Bioremediation

"Remediate" means to solve a problem, and "bioremediate" means to use biological organisms to solve an environmental problem such as contaminated soil or groundwater.

In a non-polluted environment, bacteria, fungi, protists, and other microorganisms are constantly at work breaking down organic matter. What would occur if an organic pollutant such as oil contaminated this environment? Some of the microorganisms would die, while others capable of eating the organic pollution would survive. Bioremediation works by providing these pollution-eating organisms with fertilizer, oxygen, and other conditions that encourage their rapid growth. These organisms would then be able to break down the organic pollutant at a correspondingly faster rate. In fact, bioremediation is often used to help clean up oil spills.

Bioremediation of a contaminated site typically works in one of two ways. In the case described above, ways are found to enhance the growth of whatever pollution-eating microbes might already be living at the contaminated site. In the second, less common case, specialized microbes are added to degrade the contaminants.

Bioremediation provides a good cleanup strategy for some types of pollution, but as you might expect, it will not work for all. Bioremediation may not provide a feasible strategy at sites with high concentrations of chemicals that are toxic to most microorganisms. These chemicals include metals such as cadmium or lead, and salts such as sodium chloride.

Nonetheless, bioremediation provides a technique for cleaning up pollution by enhancing the same biodegradation processes that occur in nature. Depending on the site and its contaminants, bioremediation may be safer and less expensive than alternative solutions such as incineration or landfilling of the contaminated materials. It also has the advantage of treating the contamination in place so that large quantities of soil, sediment or water do not have to be dug up or pumped out of the ground for treatment.

Bioremediation is a process that can destroy pollutants by microbial action. Certain bacteria can use nutrients and oxygen to degrade volatile organic compounds and toxic compounds in soils. In the 1980s, bans on disposal of hazardous waste at landfills led to the increased primary use of disposal alternatives such as bioremediation. However, little information has been available on the airborne emissions associated with these processes. This project was designed to help determine whether airborne emissions from bioremediation operations could pose an air pollution problem.

Decay and Renewal

Developed by the Cornell Environmental Inquiry Progammeme, Decay and Renewal consists of a student edition and teacher's guide designed to enable high school students to carry out authentic research. By using scientific protocols to investigate the processes in wastewater treatment, composting, landfilling, and bioremediation of contaminated sites, Decay and Renewal teaches important science concepts within the context of environmental issues.

Decay and Renewal includes:

- An overview of basic principles of biodegradation, the assortment of biological processes that cause organic matter to decay,
- Simple but authentic research protocols that measure biodegration in compost, landfills, wastewater treatment facilities, bioremediation sites, and streams,
- A progression of worksheets that guides students through each step of research, providing structure but flexibility in designing and conducting meaningful experiments, and
- Guidelines for integrating peer review and other collabourative knowledge-building into classroom science

Bacterial Biodegradation of High Molecular Weight Polycyclic Aromatic Hydrocarbons

Polycyclic aromatic hydrocarbons (PAH) are among the most common pollutants found at contaminated industrial sites, and in fact around the world. Although PAH are naturally occurring compounds and essentially ubiquitous at low concentrations in the terrestrial environment, high concentrations of PAH are found in contaminated soils at wood treating facilities, at sites formerly used to produce manufactured gas, and at petroleum processing operations. There are a wide range of PAH, from the simple two-ring compound naphthalene to large multi-ring compounds. We arbitrarily define those PAH with two or three rings as low molecular weight (LMW) species, and those with four or more rings as high molecular weight (HMW) species. One of the characteristic features of PAH contamination is that there is always a complex mixture of PAH, often in conjunction with a variety of other hazardous chemicals.

Environmental Protection Agency currently regulates 16 PAH compounds as priority pollutants in water, and generally considers these same compounds as "total PAH" (tPAH) in contaminated soils. The 16 regulated PAH comprise both low and high molecular weight species, and seven of them are designated as known human carcinogens. All of the carcinogenic PAH (cPAH) are high molecular weight compounds.

In addition to their toxicity, PAH as a class are extremely hydrophobic chemicals. Naphthalene is among the most water-soluble of the PAH, and its solubility in water is only about 30 mg/L. The situation becomes much worse with increasing molecular weight: chrysene (a four-ring PAH) and benzo[a]pyrene (a five-ring compound) are soluble in water in the low part per billion (mg/L) range. By comparison benzene, normally considered to be a water-immiscible chemical, has an aqueous solubility of about 2,000 mg/L. Despite the very low solubility of PAH in water, PAH-contaminated materials have been shown to be toxic in a variety of bioassays.

Microbiologists have known for a long time that microorganisms can act on most of the PAH of concern. For this reason, PAH-contaminated soils and sediments are generally considered to be candidates for bioremediation. In many of the studies in which PAH degradation in contaminated soil or sediment has been studied, however, the higher molecular weight compounds have not been removed completely. We generally do not know whether this apparent recalcitrance is due to inherently greater difficulty in metabolism of these compounds by microorganisms, a lack of bioavailability caused by their low solubility in water, or both. It is also not usually known whether microbial activity leads to complete mineralization (conversion to carbon dioxide and water) of individual PAH or only partial transformation to other organic products. Some fungi, will metabolize PAH only partially in pathways that seem to be analogous to mammalian detoxification mechanisms. While fungi do not appear to use PAH as primary carbon sources for growth, bacteria have been known for many years to be able to use several of the low molecular weight PAH as growth substrates; in fact, some of the groundbreaking work on aerobic bacterial metabolism of aromatic hydrocarbons included studies on naphthalene metabolism.

Ironically the depth of study on bacterial metabolism of naphthalene and, to a lesser extent, phenanthrene has been accompanied by only sporadic study of the metabolism of other PAH. Until the late 1980s, it was generally believed that PAH with four or more rings could not serve as growth substrates for bacteria, and could only be degraded at the expense of other growth substrates. Since then a handful of reports have appeared in the literature describing bacterial growth on the four-ring compounds fluoranthene, pyrene, and chrysene, but there is still no evidence of bacterial growth on PAH with more than four rings. There are also several studies in which the high molecular weight PAH were observed to be mineralized in the absence of growth, either by pure cultures or in complex environmental samples such as soil or sediment. We still know relatively little, however, about the mechanisms by which the high molecular PAH are degraded by bacteria.

Our work on bacterial degradation of the high molecular weight PAH began by isolating bacteria from PAH-contaminated soils. In all, we obtained ten isolates from eight different sites and one more PAH-degrading strain from another investigator. We have used these bacteria to study various aspects of PAH biodegradation, which focus on two key issues: whether the high molecular weight PAH are degraded by the same metabolic pathways a given bacterium uses to degrade the low molecular weight PAH; and how the low aqueous solubility of these compounds influences their availability to bacteria, which is prerequisite to degradation.

Mineralization of High Molecular Weight PAH

All of the bacteria we isolated from PAH-contaminated materials were obtained by enrichment culture techniques using phenanthrene as the sole carbon source. This clearly biases our observations to only those bacteria which can grow on phenanthrene, but we had hypothesized that such bacteria might be able to degrade a range of other PAH. Our hypothesis was based on the knowledge that these organisms are always exposed to a mixture of PAH, all of which are, to varying extents, structural analogues. Over evolutionary time scales such conditions would be expected to select for microorganisms with versatile metabolic capabilities. We also knew that organisms able to degrade but not grow on the higher molecular weight compounds needed to grow on something, so the low molecular weight PAH seemed like the best choice with which to begin.

We have studied each of the 11 bacterial isolates for their ability to grow on or degrade 13 PAH, ranging from two- to five-ring compounds. Most of the bacteria have a broad range of PAH substrates they can metabolize. Of most interest in terms of maximizing the biodegradation of these compounds, however, is mineralization. So far we have studied the ability of nine of the strains to mineralize pyrene, chrysene, benzanthracene and benzopyrene (the latter three compounds are all cPAH).

First, none of the organisms was able to mineralize pyrene under the conditions used in this experiment. Second, every organism either mineralized all three of the remaining high molecular weight substrates or did not mineralize any of them. These findings suggest that pyrene metabolism may not be linked directly to the metabolism of the other PAH, and that there may be a relationship among the other three PAH.

Induction of High Molecular Weight PAH Degradation

Many heterotrophic bacteria have a broad range of organic compounds they can grow on or degrade. To conserve energy for vital growth-related processes, the enzymes needed by these bacteria to degrade a particular compound are not normally synthesized unless the compound is present in the medium. Thus, biodegradation pathways in bacteria are normally induced, or "turned on," by the presence of either the parent compound or one of the pathway intermediates. If we are to learn how to control the biodegradation of compounds such as high molecular weight PAH, we clearly must understand how the relevant enzymes are turned on. This is particularly important for those compounds that do not serve as growth substrates for bacteria, since they are not likely to be able to induce their own degradation.

The only well studied example of the induction of PAH degradation in bacteria is that of naphthalene degradation. The complete pathway for conversion of naphthalene to simple, common metabolic end products is inducible by the intermediate salicylate in the bacteria for which it has been studied. In other words, the enzymes required for naphthalene metabolism are elevated to high levels in naphthalene-degrading bacteria when salicylate is introduced into the medium. (The astute reader might ask how salicylate is formed when naphthalene is the only carbon source in the medium; the answer is that in such cases the organism produces very low, or constitutive, levels of the required enzymes at all times, which then metabolize the naphthalene to form salicylate).

We have focused our attention on the inducibility of HMW PAH degradation in one of the organisms in our collection, Pseudomonas saccharophila P15. This organism was isolated several years ago by Will Stringfellow, who also conducted in-depth studies on its physiology. Both phenanthrene and salicylate induce the mineralization of benzanthracene, chrysene, and benzopyrene by P. saccharophila.

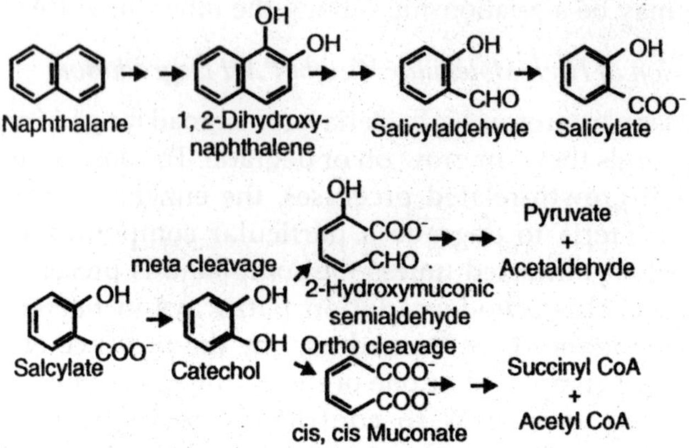

Fig.: The Principal Pathway for Acrobic Metabolism of Napthalene by Pseudomonads.

None of the three compounds are used as growth substrates by the organism, so these results suggest that the metabolism of high molecular weight PAH in P. saccharophila P15 is linked to the metabolism of a low molecular weight compound. We do not yet know, however, if the different PAH are metabolized by a common pathway or if different pathways are regulated through a common mechanism.

We hope to understand eventually if the metabolism of high molecular weight PAH by the other organisms in our collection is inducible by salicylate or other inducers of low molecular weight PAH metabolism. If we find such inducibility to be a general phenomenon, then we may have a tool with which to aid the removal of HMW PAH in bioremediation projects.

Bioavailability

Those who study the biodegradation of poorly soluble organic compounds in soil or sediment usually assume that only the amount of a chemical that is present in the aqueous phase is available to the microorganisms capable of degrading that chemical. Some microbes may have their own mechanisms of improving the bioavailability of these compounds, however. Many of the bacteria that degrade aliphatic hydrocarbons are known to produce surfactants (so-called "biosurfactants") which can solubilize or emulsify the hydrocarbons, presumably making them available for degradation at a much faster rate. Nobody knows whether there is a similar involvement of biosurfactants in the natural biodegradation of PAH. We screened our collection of bacteria for their ability to reduce the surface tension in the culture medium (an indicator of surfactant production) during growth on phenanthrene or naphthalene, but only one organism was able to do so significantly. Thus, we did not see evidence that biosurfactant production is a common feature of PAH metabolism by these bacteria. We were, however, recently able to demonstrate how synthetic surfactants can improve the rate of phenanthrene biodegradation in a well-defined experimental system

Another mechanism by which bacteria might improve their access to poorly soluble chemicals in soil or sediment is through their motility. Motility is a very common feature of soil bacteria, yet its role in biodegradation of pollutants remains virtually unexplored. Chemotaxis is a special form of motility in which an organism swims towards chemoattractants or away from chemorepellents, both cases requiring the existence of concentration gradients of the attractant or repellent for a chemotactic response to occur. The reliance on concentration gradients seems to make chemotaxis an ideal mechanism by which bacteria might be able to locate sources of "food," which in the case of hydrophobic chemicals is likely to be distributed heterogeneously in the porous medium.

We have become interested in the potential effects of motility and chemotaxis on PAH biodegradation. The lack of previous work on environmentally relevant situations has left a wide scientific gap which needs to be filled, to understand both the importance of such mechanisms in in situ biodegradation reactions and whether such mechanisms can eventually be manipulated to our advantage. If, motility and/ or chemotaxis are found to be important aspects of PAH (or other pollutant) biodegradation by soil bacteria, then we need to begin incorporating such phenomena into models that purport to simulate biodegradation processes in the subsurface.

Most of the bacteria identified are motile, and we screened most of them for chemotaxis to naphthalene, phenanthrene and known metabolites of these low molecular weight PAH. Only one of them, Pseudomonas stutzeri P16, seemed to have a chemotactic response to the chemicals of interest. Interestingly, strain appears to be chemotactic to 1-hydroxy-2-naphthoate and to salicylate, both metabolites that have been found extracellularly in media containing active PAH degraders. However, we are learning that some of the methods traditionally used to study chemotaxis are not well suited to the study of chemotaxis towards compounds of low aqueous solubility.

To test our methods with a known chemotactic organism, we recently obtained the strain Pseudomonas putida G7 from Professor Caroline. We learned that the chemotactic response of strain G7 in a standard quantitative assay becomes increasingly detectable as the starting concentration of bacterial cells is decreased.

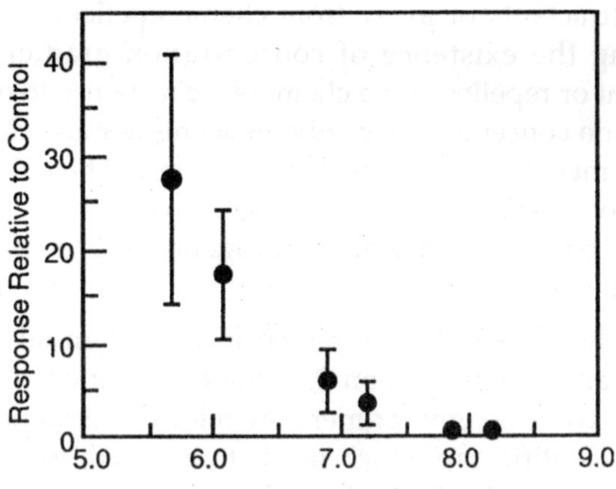

Fig.: Effect of initial Bacterial Concentration

The optimum response we observed was at a cell concentration nearly two orders of magnitude lower than the concentration typically used in the assay, which we attribute to the rapid loss of the naphthalene concentration gradient at high cell concentrations and the necessarily low starting concentration of naphthalene. Overall the response of strain G7 to naphthalene was variable yet strong. Consequently, we will focus on this organism in our subsequent efforts to understand the relevance of chemotaxis to PAH biodegradation.

There is much to be learned about the bacterial biodegradation of polycyclic aromatic hydrocarbons, and we of course can only begin to scratch the surface. In another project, we are studying whether the concepts we have explored relative to the inducibility of PAH metabolism and the effects of surfactants on PAH bioavailability can be applied to the bioremediation of PAH-contaminated soils. It is always difficult to bridge the gap between basic and applied research, but we ultimately hope that each will provide insights that will shape the future direction of our own work and that of others, both in the research community and in practice.

Pollution Prevention in Resource-related Setting

Preventing pollution in a sensitive resource-related setting means thinking through all of the activities and services associated with the facility and planning them in a way that generates less waste.

Waste prevention leads to thinking about materials in terms of reduce, reuse, recycle. The best way to prevent pollution is not to use materials that become waste problems. When such materials must be used, they should be reused onsite. Materials that cannot be directly reused should be recycled.

Everyone associated with the facility must change their habits and adopt a more responsible attitude toward waste. This includes the ownership and management of the facility - the architects, engineers, designers, and contractors; the employees; and the visitors. Each of these groups needs to consider the issues so that no waste will be generated that adversely affects the environment.

Visible, Participatory Systems

The "out of sight, out of mind" mentality regarding waste is perpetuated because the systems that deal with waste problems are all behind the scenes and off-limits. Prevention can only be accomplished by paying attention to the issue up front, rather than waiting until after the problem has been created. An environmentally sound tourism facility would ensure the visibility of systems to prevent the generation of waste. Such systems require conscious participation by visitors, users, and operators. They should not dominate the experience of a visitor at the facility. If each visitor does his or her share, the facility can be operated in a more environmentally sound manner. An additional benefit is that this can lead to long-term changes in behaviour benefiting the participant and the earth.

Training and Maintenance

Waste prevention requires training the operators, educating all users of the system, and performing diligent maintenance. Most waste problems have been created because attention has not been paid. Conventional disposal systems are designed for ease of use with the user having little idea of how the systems operate or where the waste goes. Because waste prevention represents a change in the way activities are carried out, it requires an extra effort to ensure that these practices are maintained until they become routine. In situations with high turnover of both employees and visitors, continuous training and education would be essential.

Host and Visitor Attitudes

Preventing waste means taking a measure of responsibility for activities. This can conflict with the attitude that going on vacation means escaping responsibilities. Without an attempt to change this attitude, the necessary participation may be difficult to obtain. Fostering a sense of responsibility in visitors is an important element of environmentally sound development in many other ways as well. One way to encourage this attitude is to convey to visitors that they are privileged visitors, and as a result, they may naturally feel obliged to follow certain guidelines established by their hosts. Facility operators should behave as hosts that care about making waste prevention work. Employees of the facility must care about it, and that caring must be visible to the visitors.

Garbage/Solid Waste Prevention Strategies

In planning for facilities, a comprehensive design strategy is needed for preventing generation of solid waste. A good garbage prevention strategy would require that everything brought into a facility be recycled for reuse or recycled back into the environment through biodegradation. This would mean a greater reliance on natural materials or products that are compatible with the environment.

Any resource-related development is going to have two basic sources of solid waste - materials purchased and used by the facility and those brought into the facility by visitors. The following waste prevention strategies apply to both, although different approaches will be needed for implementation:

- Use products that minimize waste and are nontoxic
- Compost or anaerobically digest biodegradable wastes
- Reuse materials onsite or collect suitable materials for offsite recycling

Use of Products that Minimize Waste

Much of the growing volume of garbage is from the use of disposable consumer products and excess packaging. Consideration must be given to materials or products that minimize waste disposal needs. Purchase items with minimal packaging; buy in bulk; replace disposable products with durable, reusable items. Durable, reusable products can be substituted for disposable ones, such as the use of returnable glass bottles instead of aluminum cans or the use of rechargeable batteries. Use of plastics for packaging is increasing, thereby replacing recyclable products and materials. Plastics, which account for about 20% of solid waste by volume, are nonbiodegradable, difficult or impossible to recycle, have a high volume to weight ratio, and are toxic when burned.

When selecting materials and goods, nothing should be purchased that will ultimately become toxic. Nontoxic materials can often be substituted for products that cause contamination problems during disposal.

Materials should be purchased locally whenever possible. Locally produced goods needing less transport and less storage should have less waste packaging.

No matter how diligent people are in purchasing materials, certain items will need to be handled. There are two methods to consider - biodegradation or recycling. Technology and economics are continually changing and the system selected must have the flexibility to adjust to market conditions. The scope of composting or anaerobic digestion of organic materials and the availability of markets for recycled materials should be considered in planning for a facility. These decisions are site-specific and must be made at the outset before final design is completed. Factors such as land availability, local markets, and isolation all contribute to that decision.

Biodegradation

In the process of biodegradation, microorganisms break down the products of other living things and incorporate them back into the ecosystem. Biodegradable or bioconvertible material includes anything that is organic. Plastics are not considered includable, despite industry contention that they are. Most of the organic components of garbage, such as paper and food wastes, can be eliminated through composting. Between 60% and 75% of the solid waste is bioconvertible.

In a tourism development, the greatest fraction of waste would be generated by the support and maintenance services. Wastes that are bioconvertible include newspapers, magazines, and wet wastepaper in the kitchen and restaurants. Also included are food waste, cardboard materials, paper office supplies, and leaves, grass, and tree trimmings. The solid fraction of the sewage waste is also considered available for the bioconversion process, and is often the most costly to dispose of otherwise, usually in a special landfill.

The biodegradable or bioconvertible percentage of the waste stream is large enough to consider at least two options for the conversion process. Two obvious options for conversion are composting and anaerobic digestion.

Composting

Composting is a familiar concept, and is used for handling yard waste and even sewage sludge. Both of these organic wastes require mixing of other materials to achieve a nutrient balance. Large chunks of relatively inert material (most commonly wood chips) add bulk and aeration to make the process work. This is typically done in open windrows or piles, with mixing done daily to provide aeration and homogeneity. This takes land space on a drainable surface, and a collection of any runoff for redispersal of the liquid to the process. It produces a quality soil amendment and reduces the bulk of the original material by approximately 40-50%.

Composting does off-gas ammonia and carbon dioxide and produces offensive orders. It typically takes 50-60 days to process. Screening of the final product is necessary to remove the bulking material, and provide granulation before use in the soil bed.

The use of this product as a soil amendment is valuable, particularly in the tropics, because the soil is essentially sterile, with only about 2% organic content. Affected by humidity, rainfall, temperature, and normal soil activity, the organic material placed on the soil will typically last only 30-40 days in the tropics. In a temperate climate, that same material may last as long as six months.

Most composting operations of the size typical for communities could be suitable for tourism development use, considering the opportunity to also process construction debris and landscape waste. This will typically be the smallest scale.

The construction debris process equipment can be transferred directly to the operations management and be capable of handling the full volume of waste.

Anaerobic Digestion

Anaerobic digestion is used extensively worldwide for the processing of food waste, animal waste, the solids of human waste systems, and for the total array of solid waste such as waste paper, green waste, and landscape waste.

This wet fermentation process converts the waste stream into three usable by-products:

- Biogas - an energy-rich gas stream, comparable to natural gas, that can be used to offset the cost of energy utilities of the development
- A high quality organic fertilizer solid that may be useful in landscaping efforts or even crop production;
- A dilute liquid organic fertilizer that may be used in drip irrigation as an additive to any planting progammeme, for foliar feeding of ornamentals, or in landscape plots for replenishing native or endangered plant species of plants.

The anaerobic digestion process does need management supervision in startup. Design of the system should be simple and modular as well as easy to operate. The modular design allows seasonal loading rather than using one large vessel. This also provides for stages of digestion that help the natural biology of the fermentation reach its optimum capacity. While this requires more capital investment, the savings to be made in maintaining operations through seasonal fluctuation will prove to be cost effective. The advantages of this system are in the smaller space it would require and in the energy it would produce, which may make the facility more self-sufficient. The value of the by-products cannot be understated, particularly in the case of remote development.

Recycling

A material doesn't become waste until it is thrown in the garbage can. If a material can be reused it is a resource, not waste. Reuse is the best form of recycling.

There are markets in the United States, Europe, Asia and other parts of the world for many recyclable materials including aluminum, paper, glass, steel, and some types of plastics. These may be available to a facility depending on location and the quantity of the materials generated. Recycling can be maximized through the purchase of products for which there is a ready market as recycled materials. The feasibility of recycling a given material depends on the price offered by a buyer for that material and shipping costs, both of which may vary over time. However, if no other reuse or recycling option exists, materials used at the facility should be recycled even at a net loss rather than sent for disposal.

In circumstances where there is no available market for a given material, often a beneficial end use can be developed locally. Every effort should be made to work with the local community to determine if any of the materials generated by the facility can be used. - e.g., glass beverage containers can be ground up and incorporated into materials for construction and road building.

Efficient recycling requires sorting of materials; convenient bins should be provided at the facility for the materials being recycled.

Offsite Disposal

If a garbage prevention strategy has been fully executed, actual remaining waste should be minimal. Remaining residuals mean that the facility is not entirely environmentally sustainable. All residuals must be collected separately and disposed of offsite. Although disposal of residual waste anywhere would have an adverse impact on the environment, in an environmentally sound development special care must be taken that this will not be borne locally. In most cases residuals must be returned to their place of origin. Toxic material residuals must be segregated and disposed of separately.

Materials Brought in and/or Purchased Onsite

Visitors and guests normally bring various materials that are consumed onsite, often leaving behind disposal problems.

Products purchased onsite and in the surrounding community can exacerbate the problem. Such materials may include toxic and difficult to dispose of items (e.g., batteries, flash bulbs, instant film refuse, and repellents. Education, including what is incorporated in the written materials about the facility and what is available, is important in minimizing these types of waste materials.

DOMESTIC WASTE MANAGEMENT STRATEGIES

General Considerations

Domestic wastes in park and tourism developments come from toilets and urinals, showers, bathroom sinks, kitchen sinks, laundry facilities, and most floor drains. Varying amounts of water are used to carry these wastes through pipes to a treatment point. Different processing techniques are used to convert the wastes for reuse or disposal. True sustainable development would not permit direct disposal of either liquids or solids before reuse.

Responsible waste management recognizes the value of reducing water needs. Properly treated wastewater can be used for toilet flushing (after approved disinfection) and irrigation (landscape needs or agricultural plantings).

The solids separated from domestic waste may be incorporated into the solid wastes (garbage) and composted into an excellent soil amendment product or anaerobically digested to produce a gaseous energy source and a residue that can also be used as a soil amendment.

If wastewater recycling for toilet flushing is used without concurrent irrigation of vegetation, there will be a volume of excess liquid that must be sensitively disposed of. This will be the amount of liquid wastes coming in each day from all sources, less the recirculated amount used for toilet flushing. Best disposal practice would use a suitable area for subsurface movement of the liquid through soils that provide good filtering and additional treatment before reaching any body of water. Direct discharge to a waterway should be the lowest priority of alternatives investigated.

Using treated, recycled wastewater for toilet flushing, rather than an equal volume of drinking-quality water, would save a large amount of water. There are also strong economic and environmental reasons for using one of more than 20 different low-volume flush toilets currently on the market, rather than the more conventional water-wasting models. In seawater development sites, toilets may be flushed with seawater and processed with septic tank and subsurface treatment and disposal. This provides a reliable system if the corrosive nature of seawater is addressed in selection of materials used in the piping, plumbing, pumping, and treatment systems.

Similar water conservation can be realised through the use of flow-restricted, fine-spray shower heads, and sink faucet aerators. Spring-loaded handles should be installed on faucets. Laundry waste volume can be reduced through the use of machines with suds-saver cycles.

Park and tourism developments that make use of water conservation devices should make an effort to educate their visitors on the amount of water (or energy) they have conserved during their stay, compared with the conventional development. It is important to monitor the conservation devices in order to evaluate their performance.

As with all pollution, sewage is best treated by not creating it in the first place. Human waste only becomes waste when there is no use for it.

Specific Waste Treatment Technologies

Conditions at sustainable developments favour simple, reliable, passive techniques for waste treatment. Systems that have few moving parts, controls, and monitoring requirements should be used. Continuity of operation and maintenance knowledge is essential. Waste treatment systems must also be able to withstand seasonal fluctuations in use. Preparation for an idle period should be simple, and no energy should be required when the system is not being used. If an innovative technology is to be employed for waste treatment, it should have a conventional backup system, or be sufficiently redundant that multiple failures are unlikely.

Dry Toilets

A composting toilet consists of a large tank located directly below the toilet room. Wastes enter the tank through a large diameter chute connecting to the toilet, and decompose in an oxygen rich environment. No water is used for the toilet, but a bulking agent (such as wood shavings) is added to improve liquid drainage and aeration, and to provide fuel. A small fan draws air through the tank and up the vent pipe to ensure adequate oxygen for decomposition and odorless operation. Internal components (such as ducts, baffles, and rotating tines) enhance the composting process. The finished compost can be removed from the lower end of the tank about once each year. It can be used as a fertilized for soil.

Composting toilets need a mild temperature, moisture, fuel, and air to function. Liquid may have to be added to the tank to keep the compost pile moist during periods of little use or a bulking agent added periodically to improve the compost texture.

Composting toilets have several advantages over other systems - e.g., no water is used and only a small amount of energy is needed for an exhaust fan; valuable nutrients are used to benefit soils; proper maintenance requires little time.

There are some disadvantages to this system. The user is close to the sewage treatment systems and this bothers some visitors. Without proper maintenance, the tank can become anaerobic and unpleasant odors arise. Undesirable pests can take up residence in the composting tank. The availability of a bulking agent in some areas can also be a problem.

Anaerobic Waste Treatment

Anaerobic waste treatment (septic systems) is accomplished through microorganisms living in the wastewater. Anaerobic microorganisms work in an environment where there is no free oxygen. Complex reactions and interactions take place with the resulting generation of some offensive gaseous by-products. These unpleasant odors are actually an indication of an efficient progression of the anaerobic process to remove the pollutants from the waste stream.

This type of system is reasonable to consider for smaller developments, where the by-product gases can be separated from occupancy areas, dissipated with good air movement, or neutralized with soil or carbon filters. Since slow treatment means longer holding periods (shallow depth tanks), large, isolated treatment and disposal areas are needed. Treated wastewater (effluent) is usually disposed of in an underground system that passes the effluent through carefully selected undisturbed soil profiles. These soils must further filter and remove nutrients as the effluent makes its way to the groundwater or other bodies of water.

One variation of this type of treatment uses part of its treated effluent for toilet flushing. The stored recycled effluent should receive some aeration to ensure odor-free recycle water in the toilets.

This type of waste treatment system has several advantages. It is relatively inexpensive to install, and is not complex to operate and maintain. It provides a consistently good quality effluent. Most components can be installed aboveground, or they can be buried or partly buried. Good quality flushing water can be provided to reduce water supply needs and final disposal volumes.

The disadvantages are that septic tank odors must be dealt with properly. Several pumps and blowers can be involved, which creates maintenance and power costs, and can cause noise in some developments. Septic tank solids require proper disposal. An aerobic digestion tank would reduce volumes for disposal. A sand drying bed can be used for digested sludge drying, and the dried sludge should be buried in a landfill. A fairly large area is required for installation. There is no known manufacturer of a complete package system.

Prior to committing time and resources, it is prudent to contact actual users of similar systems - talk with a designer, system operator, an owner, and possibly a regulatory agency inspector who has observed performance.

Aerobic Waste Treatment

Aerobic waste treatment is also accomplished with microorganisms. However in this system, air is bubbled into the treatment process to ensure plenty of free oxygen. Aerobic organisms work about 20 times faster than anaerobic organisms. No offensive gas odors are released in this treatment process. Since they process is so much faster, much less holding time is required and less treatment area is needed.

This type of treatment plant is reasonable to consider for all sizes of park and tourism development, and because offensive odors are not normally produced, this type of system can be located close to occupied areas. The normal effluent quality of a properly designed and operated system meets most all secondary treatment standards With a reasonably small filter, high quality effluent can be produced for irrigation and recycled toilet flushing water. With carefully planned effluent recycling and irrigation reuse, little or no effluent disposal would be required. If disposal is required, several options are possible.

There are several manufacturers of small aerobic treatment plants. The best plants for harsh environments and isolated areas are fabricated from durable, low maintenance materials, simple to operate and maintain, designed to use the fewest moving parts, and consistent in effluent quality. They are also quick to install and put in service, proven in similar installations, and completely assembled and warranted by one manufacturer. A good manufacturer also provides, at a reasonable cost, technical and monitoring services on the new plant for at least one year after initial startup.

Advantages of such a system are that all materials in contact with liquids are noncorrosive (plastic, fibreglass, rubber, stainless steel) and can be buried or installed aboveground. The blower is the only moving part and requires only simple, routine maintenance. Effluent quality is consistently good. Excellent recycle and reuse options are available. The process easily handles surge loadings, as well as underloaded periods. Land area required is about 0.1 acre per 10,000 gallons of rated capacity. Site installation and startup can usually be accomplished in five working days or less. Few spare parts are needed. Purchase costs are competitive with other similar-sized treatment units. Unskilled operators have successfully followed infrequent, telephoned system adjustment instructions to maintain top performance.

The disadvantages are that most components are fabricated, assembled, and shipped from Colourado - few local materials would be used. The blower, although small, needs to be run continuously. The noise would have to be muffled, and a constant power cost would be involved.

The National Park Service has five of these plants in service in four widely varying environmental areas. One plant has been in service for several years.

Alternative Disinfection Methods

Traditionally, water from conventional treatment systems is disinfected with chlorine or chlorine compounds before being released back into the environment or reused. A side effect of this is that the chlorine or chlorine compounds are very reactive and sometimes produce highly persistent, toxic chemicals. Many environmentalists believe that there is no justification for use of chlorine and its compounds for disinfection.

Most public health codes call for disinfection with chlorine, and they would have to be changed to allow for either no disinfection or the use of other disinfectants. Less desirable possibilities include using less chlorine or removing the chlorine after the proper contact time. Dechlorination requires additional chemical feed and training.

The purpose of disinfection is to ensure that no virulent organisms are present after the water has been processed in one of the systems described earlier. The most common alternative disinfectants are ozone and ultraviolet light.

An entirely new treatment technology, introduced in September 1993, appears to provide excellent disinfection without the formation of environmentally harmful by-products. The National Park Service is arranging pilot studies at existing water and sewage treatment facilities to evaluate this emerging innovative technology.

3

Life and Biogeochemical Cycles

Life on earth is inextricably linked to climate through a variety of interacting cycles and feedback loops. In recent years there has been a growing awareness of the extent to which human activities, such as deforestation and fossil fuel burning, have directly or indirectly modified the biogeochemical and physical processes involved in determining the earth's climate. These changes in atmospheric processes can disturb a variety of the ecosystem services that humanity depends upon. In addition to helping to maintain relative climate stability and a self-cleansing, oxidizing environment, these services include protection from most of the sun's harmful ultraviolet rays, mediation of runoff and evapotranspiration (which affects the quantity and quality of fresh water supplies and helps control floods and droughts), and regulation of nutrient cycling among others.

The transport and transformation of substances in the environment, through life, air, sea, land, and ice, are known collectively as biogeochemical cycles. These global cycles include the circulation of certain elements, or nutrients, upon which life and the earth's climate depend. One way that climate influences life is by regulating the flow of substances through these biogeochemical cycles, in part through atmospheric circulation. Water vapour is one such substance. It is critical for the survival and health of human beings and ecological systems and is part of the climatic state. When water vapour condenses to form clouds, more of the sun's rays are reflected back into the atmosphere, usually cooling the climate. Conversely, water vapour is also an important greenhouse gas in the atmosphere, trapping heat in the infrared part of the spectrum in the lower atmosphere.

The water or hydrologic cycle intersect with most of the other element cycles, including the cycles of carbon, nitrogen, sulfur, and phosphorus, as well as the sedimentary cycle. The processes involving each one of these elements may be strongly coupled with that of other elements, and ultimately, with important regional and global scale climatic or ecological processes. Managing and finding solutions to many of the important environmental problems facing humanity begins with understanding and integrating biogeochemical cycles and the scales at which they operate.

Examples of these links include world climate and the potential threat of global climate change; agricultural productivity and its strong reliance on climatic factors including temperature and precipitation, and on the availability of nutrients; the cleansing of toxics in soils and streams through precipitation and runoff; acid precipitation and the perturbation of ecosystem processes; the depletion of stratospheric ozone and its potential threat to human health and the food chain; and the often destructive interaction with natural cycles of other manmade compounds such as pesticides and synthetic hormones.

The Hydrologic and Sedimentary Cycles

While the total amount of water found on earth may seem huge, the amount of precipitating freshwater available to people is a tiny fraction of this total. Earth's renewable supply of water is continually distilled and distributed through the hydrological cycle. It falls from the sky as precipitation, collects in lakes, rivers and oceans or seeps into the ground, and eventually evapourates or transpires, accumulating as water vapour in clouds, ready to begin the sun powered cycle again. Water is transferred to the air from the leaves of plants primarily from a process called transpiration. This, combined with evapouration from bodies of water and the soil, is known as evapotranspiration. Evapouration of ocean water is about six times as large globally as evapotranspiration on land, although in the centres of continents evapotranspiration may be the main local source of water vapour. Changes in the global climate may cause changes in the hydrologic cycle. Increases in temperature and evapouration are expected to cause increases in precipitation, which may further affect runoff and soil moisture, and eventually influence vegetation patterns and world agriculture.

The sedimentary cycle is tied to the hydrological cycle through precipitation. Water carries materials from the land to the oceans, where they can be deposited as sediments. On a shorter time-scale, the sedimentary cycle includes the processes of physical or chemical erosion, nutrient transport, and sediment formation for which water flows are mostly responsible. On a geologically longer time-scale, the processes of sedimentation, chemical transformation, uplift, sea floor spreading, and continental drift operate. Both the hydrological and sedimentary cycles are intertwined with the distribution of the amounts and flows of six important elements - hydrogen, carbon, oxygen, nitrogen, phosphorus, and sulfur.

These elements, or macronutrients, combine in various ways to make up more than 95% of all living things. Appropriate quantities of them in proper balance and in the right places are required to sustain life. Although great stocks of all of these nutrients exist in the earth's crust in different (but not always accessible) forms, at any one time the natural supply of these vital elements is limited. Therefore, they must be recycled for life to regenerate continuously.

Phosphorus and the Phosphorus Cycle

There are a few elements that were known in ancient or even prehistoric times, examples being gold, iron, lead, and tin. The vast majority, on the other hand, have been discovered since the beginning of the modern era, and the first of them was phosphorus, which is also the first element whose discoverer is known.

A mythical substance that allegedly would turn common or base metals into gold. Convinced that he would find this substance in the human body, Brand evapourated water from a urine sample and burned the precipitate (the solid that remained) along with sand. The result was a waxy, whitish substance that glowed in the dark and reacted violently with oxygen. Brand named it phosphorus, a name derived from a Greek term meaning "light-bearer."

Owing to its high reactivity with oxygen, phosphorus is used in the production of safety matches, smoke bombs, and other incendiary devices. It is also important in various industrial applications and in fertilizers. In fact, ancient humans used phosphorus without knowing it when they fertilized their crops with animal bones.

Phosphates

In the early 1800s, chemists recognized that the critical component in bones was phosphorus, which plants use in photosynthesis—the biological conversion of energy from the Sun into chemical energy With this discovery came the realization that phosphorus would make an even more effective fertilizer when treated with sulfuric acid, which makes it soluble, or capable of being dissolved, in water. This compound, known as superphosphate, can be produced from phosphates, a type of mineral.

Phosphates represent one of the eight major classes of mineral. All phosphates contain a characteristic formation, PO_4, which is bonded to other elements or compounds— with aluminum in aluminum phosphate, or $AlPO_4$. Phosphorus fertilizer is typically calcium phosphate, known as bone ash, the most important industrial mineral produced from phosphorus. Another significant phosphate is sodium phosphate, used in dishwashing detergents. In fact, phosphates once played a much larger role in the detergent industry—with disastrous consequences

The Phosphorus Cycle

The majority of phosphorus in the earth system is located in rocks and deposits of sediment, from which it can be removed by one of three processes: weathering, the breakdown of rocks and minerals at or near the surface of Earth as the result of physical, chemical, or biological processes; leaching, the removal of soil materials that are in solution, or dissolved in water; and mining.

Phosphorus is highly reactive, meaning that it is likely to bond with other elements, and for this reason it often is found in compounds. Microorganisms absorb insoluble phosphorus compounds (ones that are incapable of being dissolved) and, through the action of acids within the microorganisms, turn them into soluble phosphates. Algae and other green plants absorb these phosphates and, in turn, are eaten by animals. When they die, the animals release the phosphates back into the soil.

As with all elements, the total amount of phosphorus on Earth stays constant, but the distribution of it does not. Some of the phosphorus passes from the geosphere into the biosphere, but the majority of it winds up in the ocean. It may find its way into sediments in shallow waters, in which case it continues to circulate, or it may be taken to the deep parts of the seas, in which case it is likely to be deposited for the long term.

Fish absorb particles of phosphorus, and thus some of the element returns to dry land through the catching and consumption of seafood. In addition, guano, or dung, from birds that live in an ocean environment (e.g., seagulls) also returns portions of phosphorus to the terrestrial environment. Nonetheless, geochemists believe that phosphorus is being transferred steadily to the ocean, from whence it is not likely to return. It is for this reason that phosphorus-based fertilizers are important, because they feed the soil with nutrients that otherwise would be continually lost.

The phosphorus cycle. Living things use phosphorus compounds in the synthesis of nucleotides, phospholipids, and phosphorylated proteins. Phosphorus enters the soil and water as phosphate ions, such as calcium phosphate, during the breakdown of crops, decaying garbage, leaf litter, and other sources.

In the phosphorus cycle, microorganisms use phosphorus in the form of calcium phosphate, magnesium phosphate, and iron phosphate. They release the phosphorus from these complexes and assimilate the phosphorus as the phosphate ion (PO_4). This ion is incorporated into DNA, RNA, and other organic compounds using phosphate, including phospholipids. When the organisms are used as foods by larger organisms, the phosphorus enters and is concentrated in the food chain.

Eutrophication

To be sure, phosphorus, in the proper quantities, is good for the environment. But as with medicine or any other beneficial substance, if a little is good, that does not mean that a lot is necessarily better. In the case of phosphorus, an overabundance of the element in the environment can lead to a phenomenon called eutrophication, a state of heightened biological productivity in a body of water. One of the leading causes of eutrophication (from a Greek term meaning "well nourished") is a high rate of nutrient input, in the form of phosphates or nitrates, a nitrogen-oxygen compound.

As a result of soil erosion, fertilizers make their way into bodies of water, as does detergent runoff in wastewater. Excessive phosphates and nitrates stimulate growth in algae and other green plants, and when these plants die, they drift to the bottom of the lake or other body of water. There, decomposers consume the remains of the plants and, in the process, also use oxygen that otherwise would be available to fish, mollusks, and other forms of life. As a result, those species die off, to be replaced by others that are more tolerant of lowered oxygen levels. Needless to say, the outcome of eutrophication is devastating to the lake's ecosystem.

As a result of high phosphate concentrations, Erie's waters were choked with plant and algae growth. Fish were unable to live in the water, the beaches reeked with the smell of decaying algae, and Erie became widely known as a "dead" body of water. This situation led to the passage of new environmental standards and pollution controls by both the

United States and Canada, whose governments acted to reduce drastically the phosphate content in fertilizers and detergents. Lake Erie proved to be an environmental success story: within a few decades the lake once again teemed with life.

Biogeochemical Cycles and Earth Systems

The changes that a particular element undergoes as it passes back and forth through the various earth systems, and particularly between living and nonliving matter, are known as biogeochemical cycles. The four earth systems involved in these cycles are the atmosphere, the biosphere (the sum of all living things as well as formerly living things that have not yet decomposed), the hydrosphere (the entirety of Earth's water except for vapour in the atmosphere), and the geosphere. The last of these spheres is defined as the upper part of Earth's continental crust, or that portion of the solid earth on which human beings live and which provides them with most of their food and natural resources.

Carbon, for instance, is present in all living things on Earth. Hence, the phrase *carbon-based life-form,* a cliché found in many an old sciencefiction movie, is actually a redundancy: *all* life-forms contain carbon. In the context of the physical sciences, *organic* refers to all substances that contain carbon, with the exception of oxides, such as carbon dioxide and carbon monoxide, and carbonates, which are found in minerals. Still, carbon circulates between the organic world and the inorganic world, as when an animal exhales carbon dioxide.

The carbon cycle is of such importance to the functioning of Earth that too is the nitrogen cycle whereby nitrogen passes between the soil, air, and biosphere as well as the hydrosphere. The hydrosphere, as noted earlier, is based on a single substance, water, created by the chemical bonding of hydrogen and oxygen, and it is likewise discussed in detail elsewhere Despite the emphasis here on carbon in the biosphere, nitrogen in the geosphere, and hydrogen and oxygen in the hydrosphere, it should be noted that biogeochemical cycles involving these four elements take them through all four "spheres." The same is true of sulfur, whose biogeochemical cycle is discussed later in this essay. On the other hand, phosphorus, also discussed later, is present in only three of Earth's systems; it plays little role in the atmosphere.

Decomposers and Detritivores

Most biogeochemical cycles involve a special type of chemical reaction known as decomposition, and for this to take place, agents of decomposition—known as decomposers and detritivores—are essential. Decomposition occurs when a compound is broken down into simpler compounds or into its constituent elements. This is achieved primarily by decomposers, organisms that obtain their energy from the chemical breakdown of dead organisms as well as from animal and plant waste products.

The principal forms of decomposer are bacteria and fungi. These creatures carry enzymes, which they secrete into the materials they consume, breaking them down chemically before taking in the products of this chemical breakdown. They thus take organic matter and render it in inorganic form, such that later it can be taken in again by plants and returned to the biosphere.

Detritivores are much more complex organisms, but their role is similar to that of decomposers. They, too, feed on waste matter, breaking this organic material down into inorganic substances that then can become available to the biosphere in the form of nutrients for plants. Examples of detritivores are earthworms and maggots. As discussed in Energy and Earth, detritivores are key players in the food web, the set of nutritional interactions—sometimes called a food chain—between living organisms.

Human Modifications of Climate Services

Human activities are significantly perturbing all of these biogeochemical cycles as well as other earth system processes, both directly through industrial processes and indirectly through changing distributions and abundance of life. The atmosphere is of particular importance to the perturbations due to its crucial role in mediating all energy that enters and leaves earth. Overall, the atmosphere is the component that controls the dominant energy flow in the earth's climate system, and solar radiation from the sun provides the energy to make the weather machine work. Embedded in this process are the biogeochemical cycles we have described that operate on a variety of time and space scales and help to regulate flows of energy and materials throughout the earth system Yet, while we understand much about the functioning of separate parts of this system, there is still a great deal to be discovered about the feedbacks and linkages that allow these interconnected parts to function as a whole and, in turn, how they will respond to human modification.

Human Disturbance

Life influences the amount of CO_2 in the atmosphere through photosynthesis, respiration, and oceanic absorption. As ecosystems are altered, the balance of these processes will be altered. Human activities are upsetting this balance and increasing CO_2 in the atmosphere through the burning of fossil fuels and clearing of forests.A significant increase in CO_2 could have dramatic consequences. Mathematical models of the climate suggest that when CO_2 (or its heat-trapping equivalent in other greenhouse gases) doubles (sometime in the middle of the next century should population, economic and technology trends continue as typically projected), the world will warm up somewhere between 1 and 5 degrees Celsius. unless other factors counteract or amplify the CO_2 - induced change Even the lower end of that range is a projected warming at the rate of one degree per hundred years, a factor of ten faster than the one degree per thousand that has been the typical average rate of natural sustained global temperature change from the end of the ice ages to warmer interglacial times. Should the higher end of the 1 to 5 degree Celsius range occur, then we could rates of climate change some fifty times faster than sustained, natural average conditions.

Climate largely determines the types of ecosystems that occupy an area. Global climate change at such a rapid rate would force many species to shift their ranges in an attempt to keep up with changing climatic conditions, as occurred during the ice age - interglacial transition ten to fifteen thousand years ago. Migrations of species such as slow growing trees with large seeds would have to occur much faster than they did in the past to keep up with rapidly shifting climates. Other species could move more easily, raising the likelihood that communities of species could be disassembled Estimating the rates of global warming in the next century, however, are very controversial because of the uncertainties involved with multiple interacting feedback mechanisms

Humanity can control climate in ways other than changing greenhouse gas concentrations. Consider the amount of moisture released to the atmosphere through transpiration in the tropical rainforests. The dense vegetation in areas such as the Amazon basin typically recycles the precipitation that falls on it many times over, helping to form heavy cloud cover in the region. The clouds, in turn, reflect sunlight and produce more rain, directly influencing regional climate as well as indirectly perturbing global climate through altering large-scale circulation patterns over the tropics. As humanity deforests regions like the Amazon, not only is CO_2 released into the atmosphere, but changes in the hydrologic cycle will almost certainly affect regional climate and possibly even global climatic patterns. In deforested areas of northeastern Brazil, the cutting of the tropical forests has led to desertification, changing both surface reflectivity and the rate of transpiration. This change in ecosystem character can lead to a destabilizing positive feedback, which may cause an even further reduction in precipitation.

In a recent study on the possible climatic impacts of tropical deforestation, researchers suggest that conversion of forest into crop land or pastures would cause significant changes in the local microclimate. Expected changes include reduction in soil moisture, larger diurnal fluctuation of surface temperature and humidity deficit, and increased surface runoff during the rainy season and decreased runoff during the dry season. Results from general circulation model simulations of large-scale deforestation and conversion to grassy vegetation in the Amazon basin indicate an increase in surface temperature, decrease in evapotranspiration, and significant reduction in precipitation. Depending on the scale of the disturbed areas, local climate changes can lead to regional climate changes which, in turn, may cause alterations in the global climate through atmospheric connections between tropical circulation and large-scale circulation patterns outside of the tropics. The effect on the ecological systems through changes in the hydrological cycle, an increase in the dry season, and the disruption of plant-animal interactions may make it difficult for the rainforests to re-establish themselves if they are destroyed. Climate change aside, the implications of this scenario for the conservation of biodiversity are serious.

The provision of fresh water and regulation of its flows through precipitation, evapouration, transpiration, and run off is mediated by all ecosystems. Forests and other vegetation types are critical components of this ecosystem service providing free flood and drought relief among other things. The loss of these services, through landuse change, can exacerbate disasters like spring floods in the Midwest and Southeast resulting from large expanses of land cleared for agriculture as well as the drainage of wetlands and swamps which otherwise might have acted as reservoirs for holding excess water or filtering toxic wastes.

Elemental Cycles that Sustain Life and Climate

Human activities are dramatically influencing the delicate balance of crucial elements in the Earth's systems, leading to changes in the complex biogeochemical cycles of carbon, phosphorus, nitrogen and sulphur. These elements are the most important for biological systems and often are limiting factors to the survival of organisms. These elements exist in a variety of forms, as gaseous compounds in the atmosphere, the cells of living organisms, or dissolved in seawater.

Components of these cycles participate at both the molecular and global levels. Crucially, however, some of the elements exist in compounds known to be major drivers of global climate change. This, in turn, has a direct impact on the geochemical cycles in which the elements participate. The entire cycle therefore responds to the levels of the compounds and hence global biogeochemistry is altered. As a consequence, the sustainability of entire terrestrial and aquatic ecosystems is potentially under threat.

Within the Alliance, an understanding of the individual biological processes at the cellular level in plants and microbes is being integrated with extensive knowledge of global biogeochemical cycles. JIC has an excellent, long-standing research reputation around biological nitrogen fixation and carbon utilisation in plants, and new research progammemes are underway focussing on the uptake of sulphur and phosphorus by plants. UEA is internationally renowned for its expertise in climate system analysis, global biogeochemical cycling and Earth System modelling.

Combining the expertise of these two institutions allows an integrated study of natural biological processes involved in elemental cycling, as well as an understanding of their roles throughout evolution. A better understanding of the natural processes involved in elemental cycling will hopefully ameliorate the effects of climate change, such as through changes to agricultural systems. This could include, carbon sequestration in soils and sediments, or more efficient use of nitrogen by crops to reduce the levels of fertilisers needed.

Soil Microbiology and Weathering

UEA research has revealed that long-term weathering processes are regulated in part by plant root exudates and the associated microbial community. In this area, the Alliance is uniquely placed to harness JIC's expertise in microbe-rhizosphere interactions to understand more fully the fundamental understanding of this process and its role within the whole Earth system.

Biological Nitrogen Fixation

The process by which atmospheric nitrogen is "fixed" into a form which plants can absorb is a major research interest within the Alliance. JIC and UEA bring significant expertise in both terrestrial and marine nitrogen fixation, and this progammeme will focus on two major themes:

- Understanding the fundamental controls on biological nitrogen fixation, and how they are affected by environmental factors (such as ambient nutrient concentration).
- Incorporating biological nitrogen fixation into global Earth System models.

Biological Trace Gas Production and Its Impacts

Certain trace gases (dimethyl sulphide, isoprene, methane, nitrous oxide and LMW halogenated gases such as methyl bromide) are significant players in elemental cycles, being involved in, ozone cycling, aerosol formation and greenhouse gas budgets. Research within ELSA is embracing the genetic, biochemical and physiological pathways that underpin production of these trace gases, linking these with biogeochemical cycling at local, ecosystem and global levels.

Climate Change Uncertainty

The combination of potentially very rapid rates of human induced climate change at the same time natural habitat has been fragmented for agriculture and development activities, and assaulted with a host of chemical agents is unprecedented. It is for these reasons that it is essential to understand not only how much climate change is likely, but just as importantly, how to characterize and analyse the value of the ecosystem services that might be disrupted. How the biosphere will respond to human-induced climate change is fraught with uncertainty. One thing that is clear is that life, biogeochemical cycles, and climate are linked components of a highly interactive system.

An illustration of this linked behaviour can be seen in the simultaneous variation of CO_2, CH_4, temperature, and SO_4^{2-} found over time in Antarctic ice cores. Temperature, CO_2, and CH_4 are positively correlated with one another, while each are negatively correlated with SO_4^{2-} . More recent data of N_2O,

CH_4, and CO_2 over the past 300 years show an increase in these trace gases that matches the magnitude of the changes in composition that occurred between the ice-age and interglacial periods. This change in composition causes more heat to be trapped near the earth's surface. Since the Industrial Revolution the build-up of these and other greenhouse gases has increased the flow of energy to earth's surface by an average of roughly two watts per square meter. Climatologists also generally agree that the global air temperature at the surface has warmed up on average approximately 0.5 +/- 0.2 degrees Celsius in the past century. It is this rate of change that appears very large compared to the sustained temperature changes from the ice ages to the interglacials in recent earth history.

Uncertainties become more significant when considering projections of climatic impacts. The combination of increasing population and increasing energy consumption per capita is expected to contribute to increasing CO_2 and sulfate emissions over the next century, but projections of the extent of the increase are very uncertain. Central estimates of emissions imply a doubling of current CO_2 concentrations by the middle of the 21st century, leading to typical projected warming ranging, from 1 degree to more than 5 degrees Celsius by the second half of the 21st century. Warming at the low end of this uncertainty range could still have significant implications for species adaptations, whereas warming of 5 degrees Celsius or more could have catastrophic effects on natural and managed ecosystems, produce serious coastal flooding, and involve other impacts on natural and human systems.

The overall cost of these impacts in "market sectors" of the economy could easily run into many tens of billions of dollars annually Although fossil fuel use contributes substantially to the cause of the impacts, associated costs are not included in the price of conventional fuels; they are externalized. Internalizing these environmental externalities into economic benefit-cost analyses is a principle goal of international climate policy advocates. We now turn to analyzing a few of the specific ecosystem services that link climate and life, and use the subjective probabilities of potential climate change impacts to provide a crude metric for assigning dollar values to certain aspects of these services.

Economic Analyses

Valuing Climate Extremes

Extreme events have become more conspicuous recently. During the summer of 1988 the Midwest experienced a record-breaking heat wave and associated drought that led to a 30% reduction in crop production that year. In early 1990's record rains led to summer catastrophic flooding of the Mississippi River and its tributaries. The winters brought severe cold spells across the country. Trends climate since the beginning of this century include a 5% increase in precipitation since 1970 over that of the previous 70 years; more than 30% of the country experiencing severe moisture surplus in each of three different years during this time period; and an average daily temperature increase of 0.3 degrees Celsius since the turn of the century.

A Climate Extremes Index produced by the National Climate Centre supports the belief that has experienced more climate extremes in recent decades, and a Greenhouse Climate Response Index shows an increase in anticipated greenhouse climate response indicators These indices combine data on weather extremes such as droughts, wet winters, severe rainstorms, and other events. While qualitatively consistent with generalized predictions for global greenhouse warming, the magnitude and persistence of these trends cannot yet be considered conclusive evidence of linkage. At the same time, however, normal variation in weather patterns may not be able to explain the increase in weather extremes since the mid-1970's, except perhaps as chance events with less than a 10% probability As part of our evaluation, we can anticipate costs associated with global change and place a preliminary value on some of the ecosystem services that could be affected.

Catastrophic floods and droughts are cautiously projected to increase in both frequency and intensity with a warmer climate as well as be influenced by human activities such as urbanization, deforestation, depleting aquifers, contaminating groundwater, and poor irrigation practices Humanity remains vulnerable to extreme weather events. This is based upon federal expenditures because private insurance losses/costs are unavailable. Ultimately, the effects of these floods are felt across a wide range of economic sectors In the Midwest flood nine states and 525 counties declared disasters.

There are numerous impacts of the flooding that are still largely unknown including cumulative effects of releases of hazardous material, including pesticides, herbicides, and other toxic materials; effects on groundwater hydrology and groundwater quality; distribution of contaminated river sediments; and alteration of forest canopy and subcanopy structure. In addition, the loss of tax revenue has not been quantified for the Midwest flood. It is important to note that while not all costs of the 1993 flood can be calculated in monetary terms, both quantifiable and non-quantifiable costs were significant in magnitude and importance. While we are not claiming that this event was directly caused by anthropogenic climate change, it does allow a rough estimate of the magnitude of costs should such changes cause increased extremes, a cautiously anticipated assessment by groups such as the IPCC

Like floods, severe droughts of the 20th century have affected both the biophysical and socioeconomic systems of many regions. Drought analyses indicate that even reasonably small changes in annual streamflows due to climatic change can have dramatic impacts on drought severity and duration. Changes in the mean annual streamflow of a region of only +/- 10% can cause changes in drought severity of between 30 to 115%. Damage estimates from the 1988 drought in the midwestern reduction in agricultural output by approximately one-third as well as billions of dollars in property damage. Hurricanes can also cause devastation in the tens of billions of dollars.Warmer surface waters in the oceans currently produce stronger hurricanes (that is, tropical cyclones are warm season phenomena). Other meteorological factors, though, are involved that may act to increase or decrease the intensity of hurricanes. An increase in intensity of hurricanes with warmer waters is plausible, yet speculative given the number of factors involved. There is little doubt, however, of the heightened damage that would be due to intense hurricanes.

Damage assessment is one possible way in which we can relate the cost of more inland and coastal floods, droughts, and hurricanes to the value of preventing the disruption of climate stability. Midwest flood example we delineated the costs of a single event. We now turn to an example of a more integrated analysis: the cost assessment of future sea level rise along coasts associated with possible ice-cap melting or with ocean warming and the resulting thermal expansion of the waters. In a probability distribution of future sea level rise by 2100, changes range from slightly negative values to a meter or more rise, with the midpoint of the distribution being approximately half a meter. A number of studies have assessed the potential economic costs of sea level rise along the developed coastline.

Valuing Carbon

There is already a historic background on the evaluation of carbon that includes climate change policies such as the introduction of carbon taxes that reduce greenhouse gas emissions through increasing prices of carbon-based fuels proportional to the amount of carbon they emit Another mechanism for reducing greenhouse gas emissions is through an international tradable emissions permit system intended to limit emissions of certain pollutants. These policies and others constitute ways of balancing the economic costs of emissions with some assumed benefit of averting the loss of ecosystem services (called "climate damage"). William Nordhaus imposed carbon taxes range from a few dollars per ton to hundreds of dollars per ton in computer model scenarios. He showed, in the context of this economic model and its assumptions, that this carbon tax would cost the world economy anywhere from less than 1 percent in gross national product to a several percent loss by the year 2100. Even a 1 percent loss in GDP, based upon his assumed baseline of 460 percent growth in personal income between 1965 and 2100, amounts to trillions of dollars per year by 2100. This cost of preventing a degrading of the climatic environment, however, should be compared to estimates of the societal value of the climate services.

Any comprehensive attempt to evaluate the societal value of climate change should include such things as loss of species diversity, loss of coastline from increasing sea level, environmentally displaced persons, and agricultural losses. Nordhaus first estimated the climate damage at 1% reduction in GNP based on market sector losses for a central estimate of climate change. This was criticized as too narrow a view of climate as a type of public good since it reflected neither non-market values (e.g. species loss) nor climate "surprise" scenarios. In response, conducted a survey of conventional economists, environmental economists, atmospheric scientists, and ecologists. Their estimates of loss of gross world product (GWP) resulting from a 3 degree Celsius warming by 2090 varied between a loss of 0 and 21% of GNP with a mean of 1.9For a 6 degree Celsius warming scenario, the respondents predicted a loss of the world economy ranging from 0.8 to 62 % with a mean estimate of 5.5%. A striking difference was noted between respondents from different academic disciplines, with natural scientists' estimates of economic impact 20 to 30 times higher than conventional economists'.However, represents climate damage of hundreds of billions of dollars annually.

While it is impossible to estimate credibly a numerical value on all of the ecosystem services provided through the maintenance of the carbon cycle at present state, it may be useful to look at land use change and loss of biomass, mostly through deforestation, as a source of atmospheric CO_2. In a very simplistic and preliminary evaluation, we can use the rates of net deforestation to calculate a value for carbon. Global loss of above-ground biomass from deforestation in the tropics is approximately 1-3 gigatons/year over the past 10 years. This amounts to between 2-5 gigatons of carbon in carbon dioxide released into the atmosphere each year from deforestation and forest degradation (this does not include the 6 gigaton carbon emissions from the burning of fossil fuels). Much of the carbon from biosphere emissions is taken up immediately by vegetation, however, leaving approximately 1-2.5 gigatons of net carbon added to the atmosphere each year. We can apply the concept of carbon taxation for emissions to an ecosystem service valuation of retaining the carbon in the forests. However, use of optimizing economic models to estimate climate damage is highly unsatisfying since these studies use very limited and often ad hoc assumptions that both over and underestimate the likely damages to various market and non-market sectors.

Methods of Valuation

The need for alternative methods of evaluation of these climate-related ecosystem services is quite clear when examining preliminary public opinion responses of global warming. In a controversial method called contingent valuation, respondents are surveyed to determine how much they would be willing to pay to prevent a given global climate change scenario from happening or accept if so much change were to be allowed. The difficulties with this type of valuing of environmental goods and processes are immense, especially since much of the evaluation is subjective. Public opinion depends, in part, upon people's exposure to the issues and the level of education and information on these issues they have received.

In a Southern California study, the contingent valuation technique was applied to the determine the influence of potential changes in temperature and precipitation resulting from global warming on respondents willingness to pay. Factorial survey methods were used to present a variety of hypothetical climate scenarios to a sample of 600 Southern California residents. Respondents were provided with a baseline microclimate for the region before future climate scenarios were evaluated. For residents living in coastal communities, the baseline climate over the past 10 years was described as having a summer average high temperature of 75 degrees Fahrenheit, with daily highs ranging between 70 and 80 degrees, and an average of 13 inches per year of rain. One possible future scenario over the next 10 years included a summer average high temperature of 100 degrees Fahrenheit, with daily highs ranging from 80 to 120 degrees (the latter typical of Death Valley, California), and an average of 20 inches per year of rain. With these and other scenarios, predicted probabilities were determined from the respondents willingness to pay for the abatement of different mean high temperatures.

This represents a 40% increment in willingness to pay for a 20 degree rise in temperature and other scenario characteristics. Note, however, that unlike the Nordhaus 1994 Survey of Experts, the Los Angeles residents reached a plateau in their willingness to pay to prevent 120 degree Fahrenheit mean temperatures as compared to 110 degree Fahrenheit mean temperatures.

However, the actual damages to the L.A. basin residents of mean high temperatures of 110 or more degrees Fahrenheit (which would imply occasional extreme heat waves similar in temperature to Death Valley mean highs) would be orders of magnitude more costly, we believe, than a 100 degree Fahrenheit mean high temperature, as such extreme heat would decimate most existing vegetation and threaten the lives of tens of thousands of elderly and other persons vulnerable to heat stroke. For just such reasons, Berk and Shulman strongly caution against taking the dollar values from the survey literally or using them in cost-benefit analyses as they confound several source of value including stewardship and altruism. In addition, some of the climate increases were well above the range of current scientific estimates of greenhouse warming.

The survey was not done in conjunction with atmospheric scientists and climatologists who could provide more realistic climate scenarios or ecologists, public health officials, or others who could help the respondents realise what such warming might mean for trees, birds, or people. Contingent valuation of the hypothetical good is possible when people believe the survey scenario. We present this type of evaluation study to highlight how difficult it is to find acceptable methods to place values on the climatic components of ecosystem services. In this survey case, the background of the respondents as well as their (limited) prior knowledge of the impacts of greenhouse warming played a large role in the survey outcomes. At the same time, however, contingent valuation points out that people are willing to pay to preserve ecosystem services as well as the tremendous need for education to help citizens more realistically value climate and other environmental

The ongoing disturbances of the atmosphere that affect the biogeochemical and physical processes that determine the climate may influence human and natural systems in profound ways. We have attempted to outline a few of the major ecosystem services that are associated with climate and the atmosphere, as well as introduce the challenging task of quantifying, and ultimately monetizing, these services.

Current monetized estimates of climate damage by the middle of the 21st century from typical climate change scenarios range from slight economic benefit to a trillion or more dollars lost annually, with most macroeconomic assessments assuming a one to two percent annual loss to GWP from climate change. Moreover, the interacting processes and biogeochemical cycles occurring across a wide spectrum of scales lead to synergistic effects that are not usually considered and sometimes not even known (i.e. surprises) when we attempt to disaggregate and value ecosystem services. The deforestation of the Amazon basin is one example of interacting scales where land use change affecting local and regional climate may also produce a net global residual. Even if the mosaic of regional effects average themselves out globally, there could be residual effects arising from heterogeneous forcing of the climate in areas outside of the tropics (i.e. regional high concentrations of sulfate aerosols or tropospheric ozone).

Ecosystems both mediate and respond to the climate system through a variety of physical, biological, and chemical feedback cycles. The uncertainty of resulting synergisms and potential global effects, as exemplified in the Amazon basin, points to the important challenge of defining and understanding the processes that link species and ecosystems with climate. With increasing knowledge, we can better anticipate ecological responses under changing climate scenarios. Meanwhile, humanity continues to perform this potentially trillion dollar unnatural experiment on "Labouratory Earth".

4

Cyclic Energy Sources

ENERGY

There is a cyclic nature to the role of energy in life and the chemicals associated with it—particularly the chemicals.

Photosynthesis and Respiration

To take you through this process quickly, notice that water and carbon dioxide can be combined together to make oxygen and glucose by a process known as photosynthesis. The oxygen is then made available in the atmosphere.

Then, through a sequence of steps, glucose can be worked into the citric acid cycle to generate carbon dioxide, water and energy. This overall process is referred to as respiration. This process depends on using oxygen made available by plants during photosynthesis. The respiration process is used by both plants and animals alike.

Glucose and Related Chemicals

Quite a number of things can happen to glucose besides being used for energy. Glucose can be changed to cellulose, to starch, to glycogen, or to other chemicals, and then, of course, back to glucose also.

Plants use the glucose to make cellulose as a structural material. Any excess glucose can be converted into starch to be saved and used later when more glucose is needed, either for additional growth, additional chemicals, or additional energy.

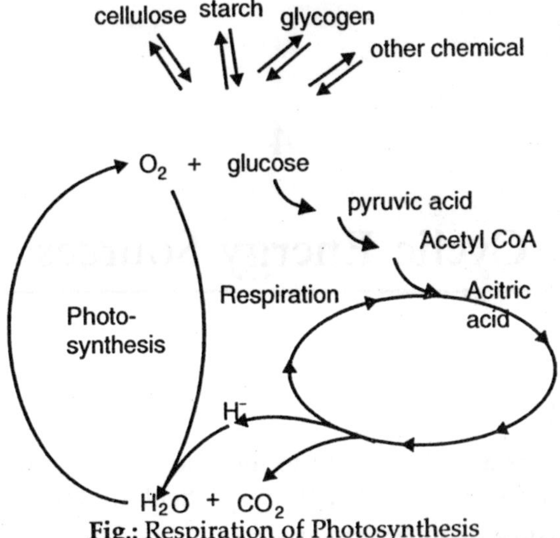

Fig.: Respiration of Photosynthesis

Animals also can use glucose to make other chemicals. We can store excess glucose in the form of glycogen for short-term future energy use. Glycogen can be broken back down into glucose to run it through the respiration process to free up some energy to carry on the functions that are needed. Glucose can also be converted into fat as long-term storage for future energy use.

In this energy overview, the glucose can then be oxidized through a sequence of processes and be changed back to water and carbon dioxide. To oversimplify the process, plants can take water and carbon dioxide and photosynthesize them into oxygen and glucose. Plants and animals both can take the glucose, change it through a sequence of chemical reactions, oxidize the glucose back to water and carbon dioxide. Sunlight drives this process. The energy that is absorbed in this process is then made available to living things (plants or animals) to drive the various kinds of chemical reactions that are necessary in a living thing.

Energy and Oxidation Aspects

One thing that you should be very much aware of is that the reaction is and endothermic reaction. It requires energy to change the reactants (water and carbon dioxide) into the products (carbohydrate and oxygen). It is solar energy that is used to drive this endothermic reaction.

Also note that if you look at the oxidation states of the elements involved there, is a change for both the carbon and the oxygen. Some oxygen ends up being oxidized, going from a -2 state to a zero state. Carbon on the other hand is reduced from the +4 state to a zero state. So there are oxidation and reduction aspects to this reaction.

Oxidation States

in H_2O	H	+1	in "CH_2O"	H	+1
	O	–2		O	–2
in CO_2	C	+4		C	0
	O	–2	in O_2	O	0

Photosynthesis can be separated into the photo part and the synthesis part.

The photo part of photosynthesis involves the oxidation of oxygen, specifically the oxygen from the water. The photo part provides us with reducing power (in the 4H·) and energy and also oxygen for animals to breathe.

The synthesis part of photosynthesis involves the reduction of carbon, specifically, it's the carbon contained in carbon dioxide. The synthesis part uses the hydrogen atoms and electrons that were generated in the photo part to reduce carbon to form carbohydrates and also water.

I should point out that both the photo and the synthesis parts of photosynthesis are multiple step reactions. Quite a number of steps are involved in separating hydrogen and electrons from oxygen in water. Similarly, quite a number of step are involved in removing oxygen from carbon dioxide and incorporating hydrogen and additional electrons into the compound. Also, the carbon dioxide is not used to create a new compound as such, but instead, the carbon is incorporated into existing carbon-containing compounds in a way that makes them larger than they were before.

The energy that's absorbed in the photosynthesis reactions can be released later (in respiration) in a sequence of steps which change the carbohydrate back to carbon dioxide and give off hydrogen and electrons, which, in turn, combine with oxygen (that we get from plants) to release energy as it's needed in the body.

Respiration

Let's take a look at some of the steps by which energy is extracted from carbohydrates and some of the reactions that are involved.

The oxygen that's in the air that we breathe in is used to combine with hydrogen and electrons to form water. The glucose that we work with goes through a sequence of steps to generate the hydrogen and the electrons which are then oxidized by the oxygen that we breathe. This process also generates the carbon dioxide that we breathe out. As the chemicals go through these steps, the carbon becomes more and more oxidized, and as it becomes oxidized, energy is released, and that energy is what is used by the body.

It is not particularly important that you know all of the steps that are involved in the oxidation of glucose, but I do think it's important that you do know that there are quite a number of steps and be familiar with what kinds reactions are involved in the process.

Photosynthesis

The main distinguishing feature of autotrophic plants is photosynthesis, and for most plants, this is a function of leaves. In this summary, I will examine the basic processes of photosynthesis, with particular concern for those elements that will absolutely critical to understanding the problems of integrated functioning of plants, or of simply making them work in your environments. We make use of the enzyme concept, as well as basic concepts about water and membranes. though in understanding biochemistry, this is relatively minor. Continuing from the groundwork that we established at the beginning of the course, this will add a physiological and biochemical perspective.

First, however, there are two important myths to dispel the gumdrop equation, and the mouse in the jar experiment. The gumdrop equation is:

$$CO_2 + H_2O \xrightarrow{h\nu} C_6H_{12}O_6 + O_2$$

Clearly the numbers don't add up as I have written it here, but it doesn't matter at all. What this equation says is that if you shine light on seltzer water, you get gumdrops (glucose). Actually, glucose is not a product of photosynthesis and probably never occurs in the chloroplasts where the process is located, and CO_2 and water don't react, even in a green bottle. So this equation simply makes a huge amount of complex biochemistry seem like the kind of thing you saw as balanced equations in 10th grade.

The bell jar experiment is the one in which some famous dead scientist dropped a mouse in a bell jar and closed it. The mouse died. He put a plant in with another mouse, and it lived. Well, actually, it died too. The reason is that even if you only use the parts of a plant that evolve oxygen, the rate is such that something like a room full of plants, continually in the light, would be needed to make the oxygen. NASA is trying this now, so is Columbia University at Biosphere II (or they were until the new president decided that only medicine mattered), and they aren't relying on a small number of plants.

Photosynthesis is the process which defines autotrophy... the ability of some organisms (i.e. plants) to grow using inorganic resources with light as the direct energy source.

Photosynthesis makes autotrophy possible because it is autocatalytic... i.e. its processes utilize substrates for reactions involved in the production of those substrates.

Eventually, photosynthesis must also be extended to include the use of the products ("sugars") either by the cells containing the chloroplasts, or other cells throughout the plants. This involves transport from the chloroplasts, or from the cytosol, or from the leaves... even long distance transport.

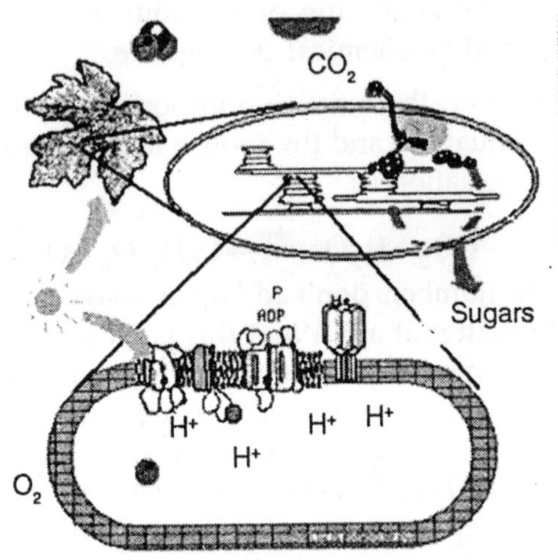

Fig.: Function of Chloroplasts

Even at the level of the inside of a single leaf, there are many things that all go on at once. If you imagine an assembly line or a fast food restaurant and think of how many things are going on simultaneously, you have the idea of "metabolic pools", at least sort of. In cells and organisms, lots of things have to be happening. Imagining a single CO_2 molecule to be processed completely before the next one can start seems pretty strange, even if it was the way your first learned about biochemistry. To understand photosynthesis, as an example of a complex but fascinating plant project, you will need to be able to understand them all together.

In the general, most common form of photosynthesis, the "Calvin cycle" is the central biochemical pathway. It is (as the name implies) a cyclic pathway of enzymatic reactions, with three identifiable parts: CO_2 fixation, C-reduction, RuBP regeneration.

In carbon fixation, a molecule of CO_2 is enzymatically attached to an acceptor molecule. The acceptor is a 5-C sugar (ribulose) with phosphates attached at each end: ribulose-1,5-bisphosphate (RuBP). The enzyme involved in the reaction is ribulose-1,5-bisphosphate carboxylase/oxygenase, which for short is called Rubisco. To give you a general idea of scales, if

CO_2 is the size of a tennis ball, RuBP is the size of maybe 10 balls, and Rubisco is the size of a house. Enzymes are big and Rubisco is a big enzyme (it has 16 subunits).

When you add a C to 5 C, you get something with 6 C. In this case, the C_6 product is unstable and spontaneously decomposes into two, 3 carbon acids. This defines C_3 photosynthesis, and C_3 is the most common form used by plants. This process uses energy and results in a lower energy product. All this means that C-fixation by itself can't be "photosynthesis".

The second step of the Calvin cycle is carbon reduction. In this step, energy is added to the product. This is not light energy, but chemical energy transferred from where it has been temporarily stored in two other sources, ATP and NADPH. In terms of ecosystems and food webs, this is the point at which energy is stored in a stable product that can be stored, or eaten. The result of these events is a 3-C product with the chemical state of sugar and an attached phosphate... called "triose-P". (There are two easily interconvertible forms, phosophoglyceraldehyde and dihydroxyacetone phosphate, in case you want to know that).

So we have the product.... the sugar. Now, it can be drained away, or it could be completely recycled to make more RuBP, or some of both. The key concept in understanding this is that of metabolic pools. Remember the mall or restaurant example: though everything happens in steps, because there are lots of people, all the steps can actually happen at the same time. In metabolism, the pools are collections of all the molecules of a certain type present in one location at one time, both enzymes and substrates. The possibilities and the choices can be summarized as follows:

The role of light in photosynthesis and its relation to the Calvin cycle. Thus far, we have been able to fix CO_2, to produce a product of the right energy content for a sugar, and to regenerate the acceptor such that as long as CO_2 can get to the chloroplasts, it can continue to be fixed. The process uses energy, however, and we have done nothing to assure that the supply is, or can be, maintained. This is where light comes in.

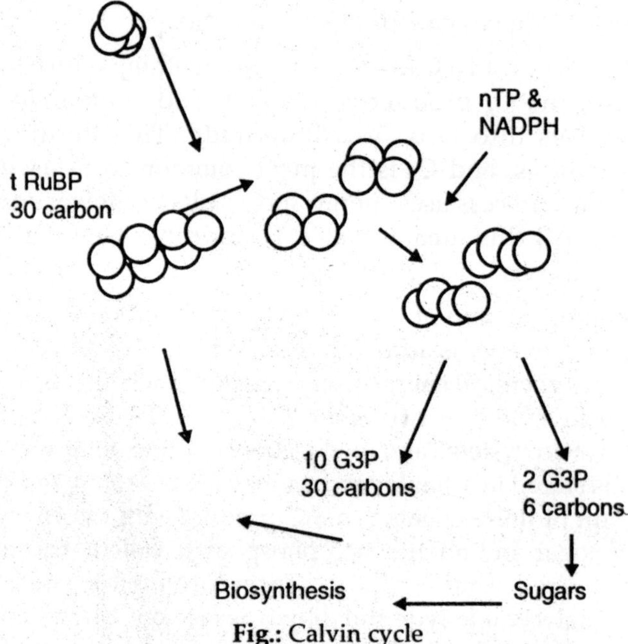

Fig.: Calvin cycle

The bottom line is that light dependent reactions produce ATP and NADPH. The chemical process uses protons and electrons from water, with the byproduct production of O_2. Splitting water is an extremely difficult thing to do. It requires a lot of energy, and to harness the energy, a multi-step, multi-protein process is involved. One very important part of this is that it occurs at a membrane. The resulting protons don't cross membranes very well. Electrons are kept within the membranes, partly because they are incredibly reactive in water. The oxygen simply diffuses through the membrane and out into the world.

ATP is made by photophosphorylation. The light dependent processes result in a H^+ gradient across the thylakoid membranes (inside acid). This gradient is a form of energy storage. The H^+ inside-to-outside gradient of potential energy would drive diffusion if the membrane were not there or were permeable to protons. As it is, the process requires a machine (enzyme) to allow the protons out. It is called the coupling factor. It allows the gradient to relax (dissipate, collapse) but in a way that uses the released energy to add a "high energy" phosphate to ADP... "coupled reaction"

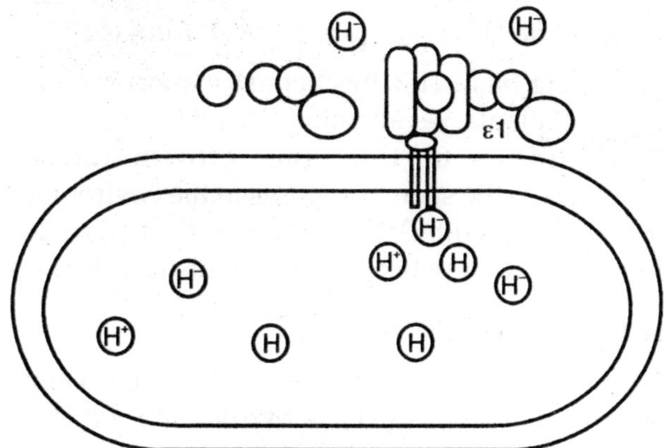

Fig.: Coupling Factor of Thylakoids

NADPH synthesis also involves the thylakoids, but it occurs at the surface of the membranes. It is catalyzed by an enzyme, of course. It requires the transfer of an electron from a donor to NADP, and NADP(H) is soluble and "free" in the stroma; it diffuses around.

Both of these processes (ATP and NADPH production) simply postpone the question of where the energy comes from.

Chlorophyll molecules in "antennae" absorb light and transfer it to a special chlorophyll in a reaction centre... e.g. P_{680}. Basically, antennae in chloroplasts work on the same general idea as antennae for radios or televisions. The absorption of the light energy by the reaction centre chlorophyll causes it to lose an electron... through an electron transport chain to another reaction centre through more electron transport chain to NADPH.

But we aren't done. Now the problem is P_{680}... it is short an electron. This electron is replaced from a protein which contains 4 Mn atoms, and which itself is now short an electron. But this can happen 4 times until the Mn cluster is short 4 electrons... this is very reactive! The role of the Mn is to store charges until there is enough oxidizing power built up to pull electrons away from water. More or less all at once, the 4 electrons are replaced by splitting two water molecules on the inside of the thylakoid membrane. This leaves 4 H^+ behind in the lumen, and releases an O_2 molecule which diffuses out. This is where the H^+ come from for ATP formation.

Concluding bits. This biochemistry and photochemistry is similar in all plants, yet all plants do not have the same rates of photosynthesis or the same growth rates (using photosynthate) or the same responses to the environment or... So, there are an incredible variety of modifications of leaf form to change the way in which the biochemistry functions.

Glucose to Pyruvic Acid

Glucose is a molecule that contains six carbon atoms, and it can be broken into parts to form pyruvic acid which contains three carbon atoms.

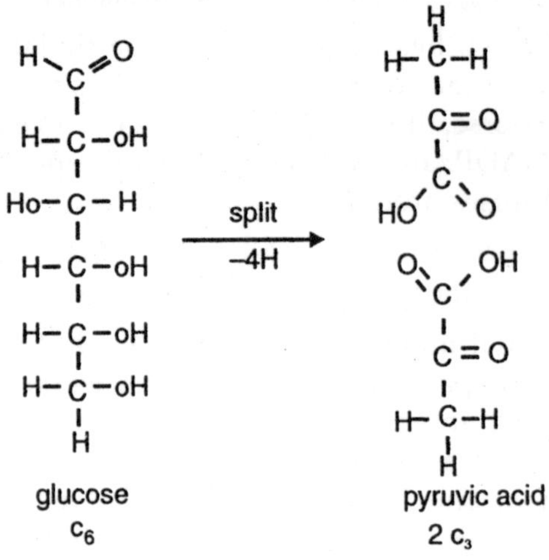

split

–4H

glucose

C_6

pyruvic acid

2 C_3

Pyruvic Acid to Acetyl CoA

One of those carbon atoms can be removed and converted into carbon dioxide changing it into an acetyl portions of a molecule which contains two carbon atoms and bonded to an enzyme called coenzyme A.

```
Pyruvic acid                                  acetyl CoA
     H                                            H
     |                                            |
  H-C-H                                        H-C-H
     |                                            |
    C=O                                          C=O + CO2 + 2H
     |       + HSCoA  ------>                     |
     C                                           S CoA
   //  \
  O     OH
```

Citric Acid Cycle and Electron Transport System

This two-carbon group is then combined with something to make a chemical called citric acid. Then there is a sequence of steps in which citric acid is successively converted into other compounds giving off carbon dioxide and hydrogen in various steps along the way.

Electron Transport System

This involves a series of oxidation-reduction reactions, in which electrons (and hydrogen) are transferred from one chemical to another to another in what is called the electron transport system.

Oxidation - Reduction

Let's return to one of the reactions in which hydrogen atoms along with their electrons are removed from a molecule. You should know from your earlier work with oxidation-reduction reactions that in a reaction like this hydrogen and the electrons don't just simply leave, they have to be removed. The reactant in this process is being oxidized. That doesn't happen by itself, there has to be an oxidizing agent.

Two oxidizing agents that operate in the citric acid cycle are simply referred to as FAD and NAD. Each one is a fairly complicated molecule and each is a nucleic acid derivative.

FAD stands for flavin adenine dinucleotide and it is synthesized in our bodies from the vitamin riboflavin. Its oxidized form is FAD and its reduced form is FADH2. Note that FAD is reduced by taking on two hydrogens and two electrons.

NAD stands for nicotinamide adenine dinucleotide and it is synthesized in our bodies from the vitamin niacin. Its oxidized form is NAD+ and its reduced form is NADH. Note that NAD+ is reduced by taking on two electrons but only one hydrogen.

NAD is the oxidizing agent that works with this reaction. Two hydrogens and two electrons are removed from the reactant (malic acid) to oxidize it into the the product (oxaloacetic acid). NAD takes both of the electrons, but it takes only one of the hydrogens. The other hydrogen, without its electron, floats around in solution as an H+ ion.

Let's again place the electron transport system in the overall context of the the oxidation of glucose and other carbohydrates. It is the portion of the process in which the hydrogen and electrons that have been stripped from the glucose and other molecules are shuffled along to an eventual combination with oxygen to make water molecules. The energy released in that process is temporarily stored (in ATP and other "high energy" compounds) for use by enzymes to catalyze endothermic biochemical reactions.

A gross oversimplification because hydrogen and electrons are removed from the breakdown products of fat molecules at many steps throughout this overall process. Electrons and hydrogen from all of those steps are run through the electron transport system to ultimately combine with oxygen and become water.

That should give you an idea of how glucose and other carbohydrates can be oxidized to provide energy and be broken down into water and carbon dioxide.

Fats for Energy

Fats can also be used to provide energy by essentially the same process as carbohydrates.

The fats can be converted into acetyl CoA and, as such, can then be brought into the citric acid cycle, be oxidized and be changed into water, carbon dioxide, and give off energy.

Proteins for Energy

Proteins can also be used to provide energy, although they serve a number of other functions as well. Burning protein to get energy is in some ways somewhat like burning your furniture to warm yourself. If you're really desperate, it makes sense. If the furniture happens to be old and broken, it makes sense. There are similar analogies to burning protein get energy. If the protein is from materials that has been broken down or is no longer functional, it makes sense to get energy out of it. Or, if you're in really serious condition, you have no carbohydrates available, you have no fats available, your body will convert protein into water and carbon dioxide and provide energy.

Energy efficiency in industry

An empirically based socio-economic analysis of successful industrial energy efficiency provides a differentiation and classification of implementation processes and resulting policy recommendations.

The paper presents the Project "Interdisciplinary Analysis of Successful Implementation of Energy Efficiency in Industry, Commerce and Service". Based on empirical case studies in Germany, Denmark, Austria and Switzerland, the interdisciplinary approach combines techno-economical variables from traditional barrier analyses with new socio-economic and socio-psychological aspects. It is the objective to gain a broader understanding of successful implementation processes in industrial enterprises.

The positive examples include energy conservation measures within the context of individual success stories as well as efficiency progammemes. The project examines the interdependencies between boundary conditions and the features and dynamics of the internal change processes analysed. It identifies main actors inside and outside of the company, the crucial determinants of their energy related behaviour, and influence factors suitable for energy policy use.

The project derives first typical patterns of social innovation and organisational development. Generalizing the findings in order to contribute to holistic policy recommendations, the project puts emphasis on instruments of economic and social marketing strategies to promote cooperative energy efficiency initiatives.

The paper presents preliminary results from the Project "Interdisciplinary Analysis of Successful Implementation of Energy Efficiency in Industry, Commerce and Service " carried out within the framework of socio-economic research of the EU JOULE progammeme. Based on the empirical investigation of positive examples of industrial energy efficiency (individual success stories and efficiency progammemes) with emphasis on small and medium enterprises in less energy intensive sectors, the paper gives a view especially from the socio-economic and socio-psychological perspective, contributing to an interdisciplinary and holistic understanding of real implementation processes.

Extended studies point at a significant share of cost effective but neglected energy efficiency potentials in industry Considering the wide range of efficient energy technologies already available on the markets, this insufficient degree of rational energy use seems to result rather from implementation deficits than from technology deficits. Especially in the case of small and medium enterprises, efficiency measures - although contributing to higher productivity, competitiveness and environmental protection - are not taken due to different hindering factors, incorporating all kind of motivation and information deficits, technical barriers, financial restrictions or market imperfections.

Regarding the discrepancy between techno-economical reasonable investment opportunities and the observed much lower degree of implementation - in literature refered to as the energy efficiency gap-the question arises, what determinants influence the investment behaviour of companies, which has been subject to empirical studies. Within the perspective of traditional barrier analysis, the company is often seen as a timeless decision making entity, reacting on economic incentives within the regulation framework and being influenced and hindered to a certain extent by barriers and obstacles, which have to be eliminated by policy intervention. An energy efficiency measure is seen as a single and detached event with its outcome determined by the related set of decision variables. This point of view reflects a rather static and one-dimensional perception of the complex issue of efficiency implementation in real business world.

Objectives and Scope

From the normative point of departure of the InterSEE project, the techno-economical approach has to be broadened by integrating and emphasising socio-psychological and socio-economic aspects because

- Efficiency measures are not stand-alone-events but part of a process
- Companies are not an entity but a complex social structure of human actors
- Company actors and their behaviour are not exclusively determined by the (rational)deliberate assessment of decision criteria but through their subjective perception of organizational culture and social reality, too.

It is the objective of the project to identify the main determinants of successful implementation processes and to gain a holistic understanding of the socio-economic interactions within and without of the companies. The findings should serve to derive typical patterns of successful implementation processes in order to structure and differentiate the target group of SME, contributing to integrated efficiency policy strategies.

Based on the wide range of analysis on insufficient implementation and related barriers, the project is concentrating explicitly on positive examples in order to identify decisive determinants and characteristics of successful energy efficiency implementation.

As a basic idealized notion, successful realization of energy efficiency means more than a single intervention reducing energy consumption for one time only. Success is characterized as a continuous improvement, i.e. a self-supporting pursuit of efficiency measures on the base of lasting behavioural and organizational change.

The project is investigating the conditions of implementation of energy efficiency in industry, commerce and service with focus on examples of small and medium entreprises in less energy intensive industries. The target group can be characterized by:

- The existence of a wide range of untapped energy efficiency potentials,
- The dominance of cross-cutting technologies, i.e. energy technologies which are not related to a specific production process like lighting, space heating etc.,
- A relatively high importance of electricity consumption
- Low financial incentives due to relatively and absolutely low energy costs
- Typical barriers due to insufficient efficiency know-how and missing personal capacities to built it up.

The research strategy of this project comprises the analysis of the implementation process within the implementing firm as well as its complex interrelations with external actors and factors of influence. In cases of a significant impact of efficiency progammemes, the analysis will be extended to the level of the progammeme actors and instruments in order to describe their impact on change, depending on the strengths and characteristic features of the particular example.

It has to be stressed, that the project does not provide a representative analysis in a statistical sense. Due to the strictly qualitative approach, emphasis is put on gaining an in-depth understanding from the contextual, process and intersubjective approach rather than carrying out a comprehensive screening of industrial activities. Taking the methodological scope for deriving general conclusions from qualitative case studies into account, the findings of the project will be able to make an useful contribution to energy policy.

Within a case study, various actors from different levels are interviewed on the base of a semi-structured interview concept, incorporating the main features of the underlying framework. The information is analysed in order to obtain an understanding of the specific implementation process observed (drawing the lifeline of success) as well as to identify core subjects, describing crucial factors of influence. Additionally, a cross-case analysis will be carried out on the base of a computer aided coding and clustering of the interview material. It is the objective to identify combinations of characteristic core subjects, specifying proto-typical patterns of successful change.

As the underlying basic framework, the distinctions between different stages of an innovation process and different levels of actors serve as the two key dimensions for pre-structuring the empirical work and preparing the analysis. The first dimension represents the development of an implementation process over time with the following stages:

- The time before the topic of energy efficiency or conservation gets on the agenda
- The arousing event (initial impulse)
- The strategic initiation of the process
- The decision phase
- The stage of concrete realization
- The project evaluation / time after the implementation (where the implementation the continuity of effects has to be ensured).

The distinction between the respective phases is ideal-typical and not suggesting that this is a straight line, rational process. Implementation can fail and stop at any of the suggested stages or fall back to earlier stages. However, the distinction provides a framework to conceptualize the investigation of trajectories of implementation processes in time in relation to the specific and often changing involvement of actors from different levels.

The second dimension concerns the different levels of actors, being more or less important for the progression of the implementation. The factors related to internal company interaction can be seperated into the level of individual actors and the level of group interaction. Concerning the individual actors, the focus should be on three different types of actors, which have different roles in the progress (in some cases certain actors may realise different roles):

- The decision-makers, who bear the core responsibility (e.g. general manager),
- The change managers, who are entrusted with the task of mediating the process of change (e.g. technical manager),
- The change agents, who take on the job of concrete realisation (e.g. technical staff),

The different actors outside the company can be distinguished with respect to the interaction with frame conditions (legal, economical, political), energy policy measures (efficiency progammemes etc.) and especially with the group of other external actors (social networks). The latter consists of similar firms, customers, energy suppliers, energy efficiency service suppliers, all kind of interest groups etc.

As depicted in Part 1 and 2, the important contribution of an integrated socio-economic approach has to be seen in providing a process understanding of energy efficiency implementation. The major findings of the first empirical case studies will be outlined and structured with respect to the following aspects:

- Efficiency measures can only be realised if efficiency is on the top of the agenda, raising the question: What makes efficiency important?
- Action not only has to be made possible, but has to be actually started: Where does the impulse come from?
- The outcomes of decisions are strongly influenced by the subjective preferences, decision criteria and decision procedures chosen by the actor: What does profitable mean in the specific context?
- Decisions have to be turned into real activities in order to accomplish an implementation process, demanding

the cooperation of other internal and external actors involved:

- Due to constraints concerning financial and personal resources as well as know-how, SME are hardly able to realise efficiency measures on their own: What is the role of external support?

Energy Efficiency as General Motivation?

The case studies indicate, that energy efficiency is rarely an objective per se. The increase of energy efficiency and the issue of CO_2 reduction seems to be too abstract, not suitable for every day identification. On the contrary, four types of motivation can be distinguished on the firm level:

General environmental concern in a broader sense (doing something good to the environment) is important for the type of actors (partially) driven by ecological philosophy. In this case, economic profitability is less source of motivation than just a necessary baseline for business activity.

- For the second type, pure economic interest in cutting down energy costs is the major motivational factor. Here, the actor is seeking for pure cost cuttings, where energy is perceived as one cost figure among others within the controlling scheme and efficiency is a techno-financial success indicator without social value.
- Thirdly, an engineering interest in process optimization especially on the technical middle management level creates general motivation for efficiency measures. These actors are seeking for technical and managerial excellence without special focus on energy, which is not a single subject but part of the general attitude of showing responsibility by "keeping the company clean". Having everything under control, i.e. technical mastership and optimization of energy consuming technologies, is seen as precondition for responsible and ethical correct enterpreneurship in a broader sense and as a proof for engineering excellence. In this case, energy efficiency is an indicator for production process productivity and quality.

- Finally, there is the actor without interest into energy efficiency at all, where energy efficiency has no priority as such.

  This is not only due to financial reasons (energy costs have a comparatively small share of total costs) but also due to time restrictions of key persons, scare personal resources are concentrated on core production issues (output, quality). In this case, the type and power of the initial impulse is of utmost importance, because a certain inertia has to be overcome (eg. by legal requirements or sharp market demands).

All these factors are not necessarily linked to branch specifications like energy intensity or exposure to environmental pressure but depend mainly on the subjective perception by the actors.

The Initial Impulse

Depending on the pre-existing general motivation, a more or less powerful kick is needed to get energy conservation measures going. The type and source and especially the right timing of the initial impulse are important parameters determining the beginning of successful implementation processes. Initial impulses can be roughly divided into:

- Production technology related requirements such as recurrent technical breakdowns, technical constraints of the local infrastructure or the occasion to link the measure with investment which has to be undertaken anyway. Especially in this case, the companies depend on internal or external impulses to realise efficiency measures within the regular investment, stressing the importance of the right kick at the right time.
- The inspiration of high and middle management are major sources of initial impulses. The latter, however, are often influenced by external impulses as through contact to experts, consultants and free or cheap energy audits.
- Environmental regulation can provide a decisive push towards ecological activity, especially as first steps for a company being non-active so far (Holm et.al. 1994). The

same holds for the case of costumer side pressure demanding clean products.

Money Talks - the Perception of Economic Parameters

In most companies cost effectiveness in the sense of economic reasonable investment is a mandatory baseline for decision, i.e. apart from single exceptions non-profitable options are not taken. The subjective perception of profitability, however, differs significantly, which is not necessarily linked to business success. On the one side, only measures fullfilling strict quantitative requirements are put into action and financial aspects are very important for the decision process.

By contrast, other companies try to execute every reasonable and useful efficiency measure evaluated in a broader qualitative sense. Energy costs and efficiency are seen only as one factor among others, connected with multiple features of the measure such as working conditions, increase in quality or productivity etc. in this case, the restriction on energy costs is seen by the executives as a destructive fixation on the financial dimension, destroying the holistic value of a multifunctional investment..

Interestingly, in quite some cases the quantitative saving effect of a conservation measure is not exactly known by the companies due to insufficient monitoring and measurement. Even in cases where costs savings are perceived to be significant and relevant to the decision, reliable data is rarely obtainable, leaving the assessment open to personal judgement. This lack of objective decision parameters underlines the subjective character of profitability assessments.

Relating Organizational Culture

Successful efficiency implementation depends on team work, because energy efficiency measures are often concerning complex technical problems, influencing different parts of the company. In order to achieve the full amount of possible savings, proper operation and permanent awareness (good housekeeping) from the staff's side is necessary.

Once decided, however, energy efficiency projects need a change manager who is taking the initiative and responsibility for the accomplishment of activity. Very often it depends on his personality, whether a project can be put through against internal reluctance and resistance. Especially with projects not directly connected to the core business or when unfamiliar solutions have to be implemented, a strong management and the leading role of innovative high hierarchy personnel are important fostering factors for the initiation as well as for the implementation and continuation of a measure. Sometimes, the commitment of the owner and top manager of the firm, is connected with an authoritarian management style, so that objections on the side of the employees are overcome through strict orders. Problems regarding the internal acceptance can turn up at the beginning of energy efficiency projects, as the involved employees are not yet fully convinced of the project, but have to face a bigger work load in the course of the realisation.

Charismatic leadership of the top management, aiming to influence the opposition of internal actors by participation from the beginning and communication refering to objective physical measurement, is one of the most important drive for successful activity. It creates a fostering organizational culture, providing a risk-free and stimulating atmosphere for activities, being essential for creating the middle management's crucial commitment to the corporate philosophy. The participation of all relevant actors from the beginning of the project contributes to less internal resistance, better results through additional ideas and inputs including experiences from the worker's side. For continuing improvement processes it is indispensable, that every member of the organisation commits himself to the common goal of higher efficiency. However, the case studies indicate that an engagement of staff grows slowly and it takes some time and communication effort to generate self-supporting participation processes. As observed in the case studies, many contributions to energy savings are stemming from waste management and quality measures, because these fields are much more concrete and easy to visualize - the energy and climate subject remains often to abstract, impeding full identification of workers with the issue.

Success Needs External Support - The Role of Social Networks

The input of external know-how is in nearly all cases of crucial importance. Energy audits, consultancy or networks provide impulses, information and concrete know-how for solutions which cannot be built up the the companies itself. In some cases, fostering network relations have been built up, where internal engagement is fostered more generally through contacts with other firms making the same experiences, providing regular exchange of information and experiences. It has to be stated, however, that certain limitations has to be depicted in relation to these cooperations. Generally, the network observed supports mainly already existing, fairly advanced movements and does not create new activities from the scratch. Networks demand engagement, representing effort and investment into social links, which can reach significant dimensions and might exceed the SME´s capacities. Furthermore, networks demand commitment and trustful cooperation, which is hindered in case of direct competition between the participants.

However, most of the restrictions and barriers can be overcome by open minded attitude of all actors involved and by sophisticated moderation, confidence and creating stimulating social learning processes.

Furthermore the case study results indicate, that access to a network of professional energy specialists in the sense of a technical hotline, providing specific information for detailed realization problems, would be an important fostering factor for the technical management.

The Dimensions of Successful Change Processes

The case study findings have pointed at a wide range of process features, some of them ambigous in impact and strength or even contradictory. Many details have been skipped due to the limitation of this paper presentation and much more could have been said about interdependencies and temporal relations or causalities. However, some aspects can be found throughout all considerations: In terms of fostering energy efficiency, actors on all levels have to develop motivation and commitment to the goal of reducing energy inputs, i.e. change has to take place on the level of the willingness to act.

Additionally, the course of implementation demands technical and social skills, cooperation with internal and external partners and access to sufficient personal, technical and know-how resources. Since in most cases SME and the staff members are lacking sufficient expertise in at least one of the areas, change has to take as well on the level of the technical and organizational capability to act.

Consequently, the scope for companies to increase energy efficiency is determined by the pre-existing state of development, i.e. the history of efficiency in the company with regard to the general motivation to tackle that issue and the existing organizational and technical competence. Within the course of an ongoing improvement process, the company can develop with respect to these three dimensions, and different trajectories are possible.

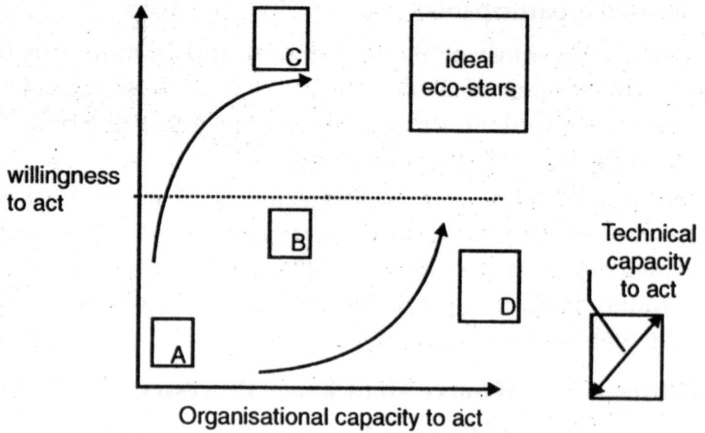

Fig.: Three Dimensions of Succesful Change Process and a Proto-typical Classification of Firms

In relation to the internal capacities and the external impulses or inputs, the company starts to built up motivation, technical and organizational competence. The underlying steps differ for instance in terms of actors involved, time or money required, internal resistance, external support or opportunities for cooperation.

TYPE A: This type has neither sufficient motivation nor competence to enter a successful self-dynamic implementation process. These companies depend on external kicks to get going such as regulation, financial incentives or strong social inputs. Even if positive experiences are already made, energy conservation measures are only taken isolated without continuation. In spite of insufficient activity on the firm level, some factors might be fairly advanced, e.g. concerning the engagement of the technical manager, who still stands alone with his ideas. These advanced factors represent the first staring points for energy policy, which have to be identified and addressed.

TYPE B: This type shows a certain willingness to act, but has no sufficient know-how to act on its own. To certain extent, however, enough organizational competence could have been developed in order to use and exploit external inputs, but without deliberately and systematically seeking for. This company implements nearly every measure being told by consultants or other external actors but does not reach the state of a self-supporting process yet.

TYPE C: This type is characterized by very high motivation and fairly good techncial and organizational competence. Energy efficiency and environmental concern are parts of the corporate philosophy and most staff members show strong commitment to these common goals. Since in this case motivation and enthousiasm are higher than the necessary implementation skills, this firm is facing the risk to overestimate its possibilities by tackling unrealistic tasks. Sometimes, this leads to (unnecessary) failures, putting a damper on the actors' engagement.

TYPE D: This type is characterized by the development of sufficient technical and organizational resources, driven by production or cost reasons. Often the staff has first elabourated technical know-how before gaining additional organizational skills by management systems, controlling etc. The company might even be very open minded in terms of innovation and adaption of advanced technologies without having a corresponding ecological philosophy. Due to this high competence, positive experiences can be expected for first efficiency measures contributing to stimulation by feedback of success.

Even if not every company will fit into this structure, the classification presented gives a first glance on different types and stages of development processes. The main features described will have a strong influence on the character and pace of change processes. Especially the aspect of process speed will have to be taken into account by effective energy policy strategies in order to place the right incentive at the right time:

Role of Process Pace

The realization of energy conservation projects demands financial and personal resources - and most SME are lacking both of them. Additionally, the technical competence, especially concerning non-core activities, is limited and cannot be easily increased. Even external support needs an internal contact being subject to the same limitations. All these restricting factors indictate, that SME have to take small and manageable steps in order to achieve their efficiency goals. Furthermore, small steps reduce financial committment and thus reduce risk of wrong decisions. Accordingly, in most cases, success is not characterized by the optimal turn-key solution but by a permanent step-wise approach followed over longer time. From the socio-psychological perspective, this approach is one of the most important preconditions for successful and lasting processes. Every step provides results, which can be feed into an internal feedback loop. The feedback of positive experiences is a major source of motivation, confidence, trust and readiness to take risk, increasing the self-efficiency of engagement. Instruments like energy management systems and internal communication channels can contribute to this feedback by providing measurements and data, and by documenting and multipliying the achievements.

A Broader Understanding of Innovation - The Need for Social Innovation

Successful change processes increasing energy efficiency in SME depend on new and unfamiliar devices, methods and behaviour, i.e. on innovation. Within the context of the project and regarding the company typology derived from the case studies, innovation can no longer be exclusively interpretated as a product-oriented improvement of the technical features of energy technologies. Without doubting the need for further R&D in the technical field, the case studies indicate, that the energy political focus has to be shifted more towards process-oriented innovations. The results of the case studies and the derived three-dimensionality of change processes (motivation, technical capability, organisational capability) emphasise the importance of social innovation. Social innovations increase the willingness to act and organizational capacities to act. They have to be introduced by internal activities such as new communicative structures (e.g. efficiency work groups), management systems or the documentation and publication of positive results. At the same time, social innovations refer to external links such as local or branch-specific networks, social and political engagement and customer-supplier relations. As any innovation, social innovations cannot be commanded by politics but a stringent policy framework can create a fostering environment, making it more likely to spread.

The case study results stress once more the existence and importance of energy efficiency potentials on the industrial end user side. Within the target group analysed, energy conservation measures contribute to a significant better performance of the companies by reduced energy costs, better working conditions, motivation of staff, higher quality and productivity or powerful management systems.

In order to address these untapped potentials, the findings from the case studies and the derived socio-economic perspective on the implementation process can contribute to a modification of traditional, rather linear and static understanding of energy policy. Energy efficiency should not longer only be seen as result of techno-economical equilibrium finding, partially affected by timeless barriers and market imperfections. By contrast, the findings point at some aspects generating an intergrated view on dynamic interaction, leading to an integrated implementation support and social marketing strategy:

Policy Mixes for Implementation Support

Integrated policy mixes providing support and assistance through the whole implementation process from the very beginning to the end will contribute to increased effectiveness and efficiency of policy strategies. That does not mean, that measures are entirely initiated and realised by public means and funds. By contrast, the focus should be on initiating a self-supporting change process within the company, which has to be strengthened in decisive moments by a fostering environment. This demands a specification of the instrumental bundles with correspondance to the target group and stage of the process. Additionally, the schedule and evaluation procedures of policy strategies have to take the small steps and thus relatively slow pace of change processes into account. Main features of a process orientied implementation support strategy are:

- Mandatory differentiation target group: The specific definition of target groups and how to address them suitably according to their prevailing motivation is a major challenge for policy makers. The average energy intensity of a branch or single technologies are likely to be a misleading indicator in setting target group priorities for energy conservation policy. Special analyses are necessary to identify more precisely these firms that consider energy consumption a promising field for action in order to define target groups more aptly.

- Differentiation of economic criteria: implementation results are not exclusively determined by economic critieria. Even if cost awareness contributes to general motivation, the impact of cost-benefit calculations is often restricted to the pure decision phase. Very important factors bringing efficiency on the agenda (pushing a decision) and ensuring the realization success (implementating a decision) are to a high degree influenced by social parameters, demanding non-economic instruments.
- Increasing internal competence by external support and social networks: The internal competence and thus capability to initiate and realise efficiency measures has to be increased by various means:
  - Supply of practice-oriented technical know-how and information by special training and education for technical personal, planners etc., and network contacts such as a technical hotline, providing precise information to detailed implementation problems.
  - Supply of management know-how by introduction and support of energy and evironmental management systems and decision tools.
  - Supply of organizational and social know-how by information and training on organizational development, methods and instruments of internal communication and cooperation and staff participation

Social Marketing

Energy policy should put more emphasis on the initiation of self-enforcing processes in order to achieve a continuity of engagement and activity within the target group even without further policy interference. Key players, such as inspired executives and decision makers especially from the high management level, have to be activated and supported. In order to raise their concern and motivation for the issues, special activities to address this target group despite its severe time restrictions are essential. Thus, it is necessary to evaluate the outcome of measures and to provide a feedback of success by actors specific marketing and communication strategies. Pilot studies or demonstration projects of best practise in typical applications are suitable instruments to stimulate initiation and motivation if they are combined with an adequate marketing strategy e.g. by decentrally organized workshops. Competitions and awards provide a platform to give energy saving actors a positive feedback on their efforts and thus contribute to an increased motivation.

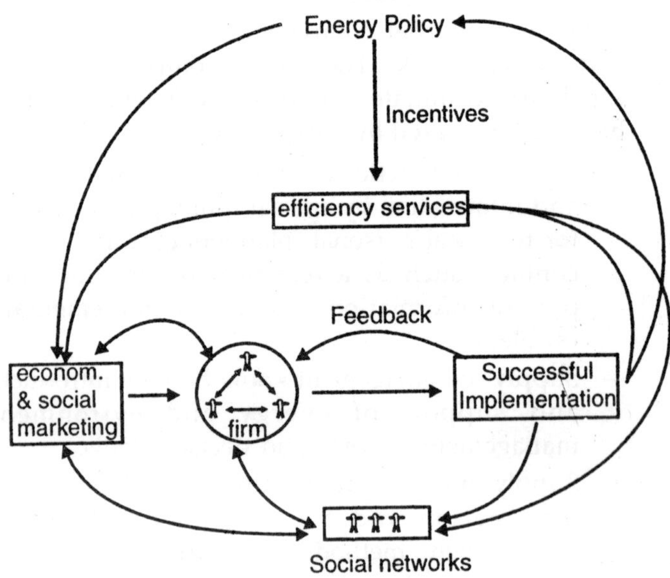

Fig. Cyclic and self-enforcing Energy Policy Approach

And finally: policy makers too should look upon energy efficiency measures as "products" which have to be "sold" to the target group of SMEs. Beside a appropriate differentiation of the targeted actors, policy strategies and instruments have to be accompanied by sufficient marketing and communication efforts. Actors within companies and institutions have to be adressed and convinced by throughful selected and designed marketing instruments, requiring a sufficient share of funds for this purpose.

5

Ecosystem and Biogeochemical Cycle

ECOSYSTEM

An ecosystem is a group of living and non-living components interacting together on a given physical landscape. The size of an ecosystem is arbitrary and could be as small as a few square centimeters if you are looking at a soil microbial ecosystem; as large as thousands of square kilometers if you are looking a biome like the Great Plains ecosystem; or a few hectares if you are looking at a single forest stand ecosystem.

FORESTED ECOSYSTEM WORKS IS TO BUILD A MODEL

An ecosystem model is an accurate but simplified representation of an ecosystem that can be very useful in thinking about or simulating the actions of a real ecosystem. Because any ecosystem has many different but interrelated components, the best way to understand the system is to break it down into its component parts. To get an introduction to a very simplified forest model, which gives participants and introduction to how a hardwood forest ecosystem works before and after exotic earthworms invade!

Step One: The first step in building a graphical model of a hardwood forest ecosystem is to identify its major components.

The components of any ecosystem are those physical things that contain energy and nutrients.

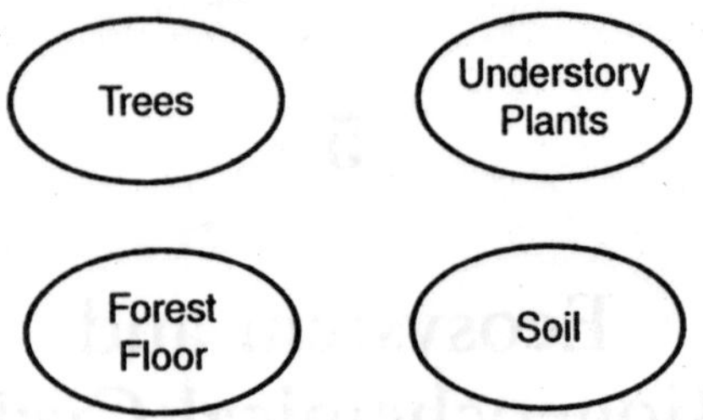

Fig.: Components of a Graphical Forest Ecosystem

A forested ecosystem, by definition contains trees, so that is our first component. In addition to the various species and layers of trees in a forest, there are other distinct ecosystem components. The *understory* contains most of the visible plant life found between the sapling layer and forest floor. The forest floor is where one would find most of the plant roots, bulbs, fungi, seeds, years of accumulated leaves and twigs. The soil is the "dirt" under the forest floor and is composed largely of minerals of various grain size (very small grain size = clay…very large grain size = sand) and organic material that has been mixed in with the mineral component. In addition, there are numerous animals that live in the forest, and of course we cannot forget people. We will add those two components to our forest ecosystem model later

Step Two: Once you have identified the components of your ecosystem model, you need to define the *processes* that connect the components. This is graphically done by using arrows to indicate the flow of nutrients or energy among the different ecosystem components

The components of our ecosystem model are now connected by processes that result in the movement of energy or nutrients among the components.

One thing to notice in our ecosystem model is that there is no process connecting the *trees* component with the understory plants component. This is because there are no substantial processes that result in the flow of nutrients directly from a tree to an understory plant or visa-versa. There would be important relationships between the trees and understory plants. Trees provide shade to the understory plants. But remember, in an ecosystem model, only processes that result in flow of energy or nutrients are represented. Now let's add the animals and the people components to our ecosystem. energy & nutrients flow from the trees and understory plants to the animals when they eat the leaves, twigs and buds of trees or graze on understory plants; and when the animal excrete waste products or die, energy & nutrients are returned to the forest floor component. Since *people* are really just a special kind of animal, energy & nutrients flow from the trees to people when they eat something from a tree, like maple syrup

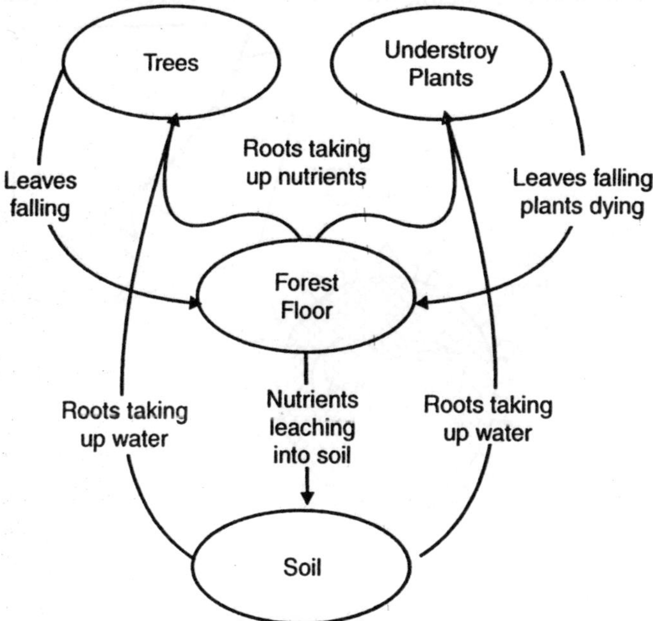

Fig.: Movement of Energy or Nutrients Among the Components.

We have added two components to our ecosystem model, along with some processes connecting them to other components.

Step Three: Determine the major inputs and outputs of your ecosystem. As you are building your ecosystem model, one thing to think about is whether your ecosystems could be opened or closed. A closed ecosystem is one that has no inputs of energy or nutrients from outside the ecosystem and no outputs of energy or nutrients leaving the system. The earth is an example of a closed ecosystem with respect to nutrients and an open ecosystem with respect to energy All the nutrients that have ever been on earth are here and simply continue to cycle, there are no additions or losses. However, the earth is constantly getting inputs of energy from the sun and simultaneously radiating energy back. The earth doesn't heat up too much or cool down too much because the earth's energy balance is in a relatively stable equilibrium, meaning that the amount of energy being input and output are about equal.

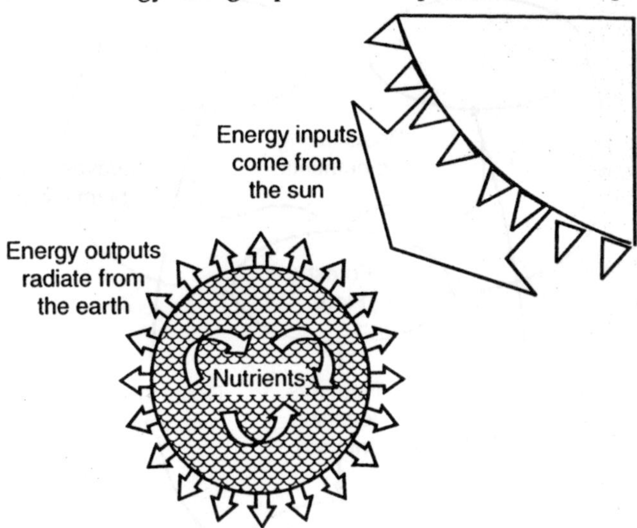

Fig.: The Earth Ecosystem

The earth ecosystem and has no inputs or outputs of nutrients which are constantly recycled within the global ecosystem, while the earth has both inputs and outputs of energy that are in a relatively stable equilibrium.

Just as the Earth ecosystem is closed with respect to nutrients, unmanaged earthworm-free hardwood forest ecosystems are often very nearly closed nutrient ecosystems that there are very few inputs or outputs of nutrients. Rather the nutrients are constantly recycled among the various ecosystem components. In contrast, most agricultural ecosystems require nutrient inputs from outside to function properly

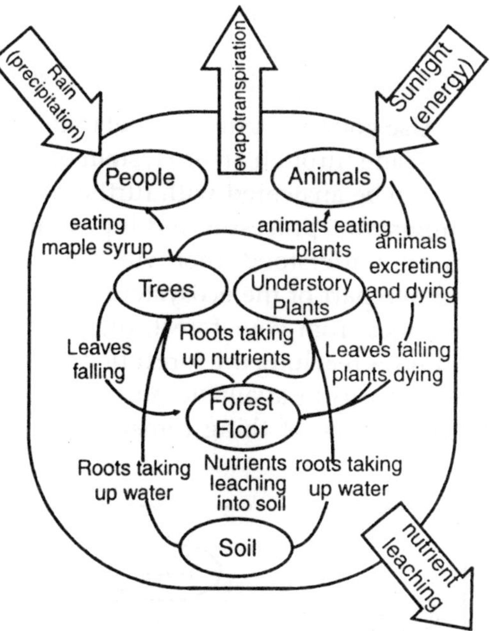

Fig.: Evapotranspiration

Some typical inputs and outputs of nutrients and energy for forested ecosystems include evapotranspiration, nutrient leaching, sunlight and rain.

Step Four: Once you have identified the components, processes and major inputs and outputs in your ecosystem model, then you can begin to add the actual values to these parts of your ecosystem by measuring them. The amount of litter that falls to the forest floor each year (a process), what the biomass of trees is in a given forest (a component), how much light reaches the forest over a growing season (an input), or how much nitrogen leaches from the forest (an output). Needless to say, some of these things are easier to measure than others and for most of these things it would be very hard to directly measure the value for a whole forest. It would be hard to catch every single leaf that fell from the trees in a given year and weigh them all! So, researchers estimate these values taking samples of the given measurement they want to know. In the case of leaf litter, you can put out trays in the forest and after all the leaves have fallen for the year, dry and weight the leaf liter in your trays. They you can use that value to calculate an estimate of the total leaf litter for your forest

Step Five: Use your ecosystem model to think about how changes can cascade through an ecosystem or to ask specific questions that can be answered with further research. When the major components, processes and inputs and outputs of an ecosystem are understood, then you can use the model how changing one part of the ecosystem affects other parts. If you harvest trees from your forest, that will decrease the amount of leaf litter reaching the forest floor each year which may lead to decreases in available nutrients for understory plants. This is the type of thing forest ecology researchers often study.

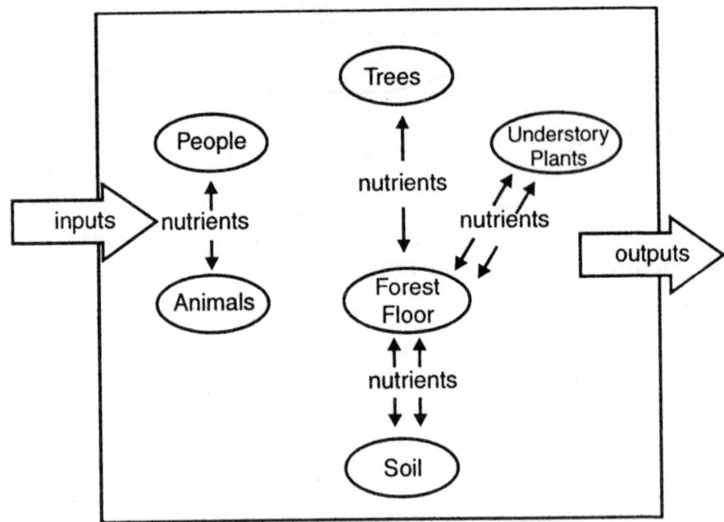

Fig.: Hard wood forest Ecosystem Model.

An ecosystem that is in equilibrium doesn't gain or lose nutrients.

Researchers may monitor soil nutrient levels for many years after trees have been harvested how the real forest behaves compared to what they thought might happen based on their forest model, their understanding of how the forest works. If the results in the real forest are very different than those predicted by their model, then they know that they don't have full understanding of how their forest works and they may go back to try to improve their model.

In general, when building a model, you start simple and add more detail as needed. To get an introduction to a very simplified forest model which gives participants and introduction to how a hardwood forest ecosystem works before and after exotic earthworms invade!

ENERGY FLOW THROUGH THE ECOSYSTEM

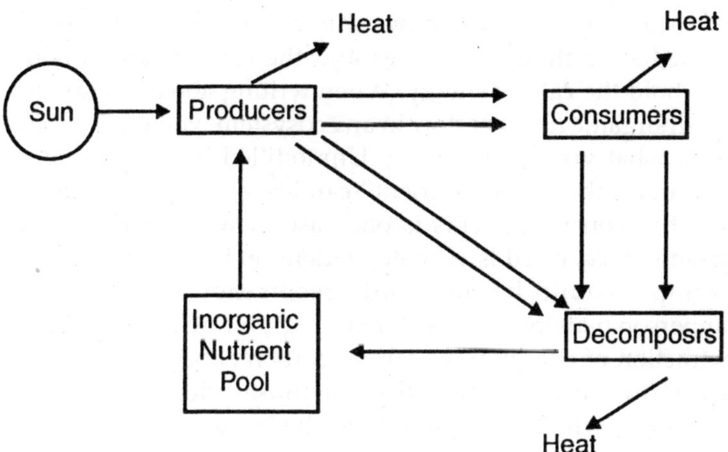

Fig.: Energy and Inorganic Nutrients flow through the Ecosystem

Both energy and inorganic nutrients flow through the ecosystem. We need to define some terminology first. Energy "flows" through the ecosystem in the form of carbon-carbon bonds. When respiration occurs, the carbon-carbon bonds are broken and the carbon is combined with oxygen to form carbon dioxide. This process releases the energy, which is either used by the organism or the energy may be lost as heat. The dark arrows represent the movement of this energy.

The other component are the inorganic nutrients. They are inorganic because they do not contain carbon-carbon bonds. These inorganic nutrients include the phosphorous in your teeth, bones, and cellular membranes; the nitrogen in your amino acids; and the iron in your blood. The movement of the inorganic nutrients is represented by the open arrows. Note that the autotrophs obtain these inorganic nutrients from the inorganic nutrient pool, which is usually the soil or water surrounding the plants or algae. These inorganic nutrients are passed from organism to organism as one organism is consumed by another. Ultimately, all organisms die and become detritus, food for the decomposers. At this stage, the last of the energy is extracted and the inorganic nutrients are returned to the soil or water to be taken up again. The inorganic nutrients are recycled, the energy is not.

Many of us, when we hear the word "nutrient" immediately think of calories and the carbon-carbon bonds that hold the caloric energy. When writing about energy flow and inorganic nutrient flow in an ecosystem, you must be clear as to what you are referring. Unmodified by "inorganic" or "organic", the word "nutrient" can leave your reader unsure of what you mean. This is one case in which the scientific meaning of a word is very dependent on its context. Another example would be the word "respiration", which to the layperson usually refers to "breathing", but which means "the extraction of energy from carbon-carbon bonds at the cellular level" to most scientists (except those scientists studying breathing, who use respiration in the lay sense).

To summarize: In the flow of energy and inorganic nutrients through the ecosystem, a few generalizations can be made:

- The ultimate source of energy (for most ecosystems) is the sun
- The ultimate fate of energy in ecosystems is for it to be lost as heat.
- Energy and nutrients are passed from organism to organism through the food chain as one organism eats another.

- Decomposers remove the last energy from the remains of organisms.
- Inorganic nutrients are cycled, energy is not.

FOOD CHAINS AND WEBS

A food chain is the path of food from a given final consumer back to a producer. For instance, a typical food chain in a field ecosystem might be:

grass → grasshopper → mouse → snake → hawk

Note that even though I said the food chain is the path of food from a given final consumer back to a producer we typically list a food chain from producer on the left to final consumer on the right. Note to international readers: In Hebrew or Aramaic, or other languages which are read right-to-left, is it customary to list the food chains in the reverse order? By the way, you should be able to look at the food chain above and identify the autotrophs and heterotrophs, and classify each as a herbivore, carnivore, etc. You should also be able to determine that the hawk is a quaternary consumer.

The real world, of course, is more complicated than a simple food chain. While many organisms do specialize in their diets (anteaters come to mind as a specialist), other organisms do not. Hawks don't limit their diets to snakes, snakes eat things other than mice, mice eat grass as well as grasshoppers, and so on. A more realistic depiction of who eats whom is called a food web.

It is when we have a picture of a food web in front of us that the definition of food chain makes more sense. We can now a food web consists of interlocking food chains, and that the only way to untangle the chains is to trace *back* along a given food chain to its source.

The food webs are *grazing food chains* since at their base are producers which the herbivores then graze on. While grazing food chains are important, in nature they are outnumbered by *detritus-based food chains*. In detritus-based food chains, decomposers are at the base of the food chain, and sustain the carnivores which feed on them. In terms of the weight (or biomass) of animals in many ecosystems, more of their body mass can be traced back to detritus than to living producers.

Pyramids

The concept of biomass is important. It is a general principle that the further removed a trophic level is from its source (detritus or producer), the less biomass it will contain (biomass here would refer to the combined weight of all the organisms in the trophic level). This reduction in biomass occurs for several reasons:

- Not everything in the lower levels gets eaten
- Not everything that is eaten is digested
- Energy is always being lost as heat

It is important to remember that the decrease in number is best detected in terms or biomass. Numbers of organisms are unreliable in this case because of the great variation in the biomass of *individual* organisms. For instance, squirrels feed on acorns. The oak trees in a forest will always outnumber the squirrels in terms of combined weight, but there may actually be more squirrels than oak trees. Remember that an individual oak tree is huge, weighing thousands of kilograms, while an individual squirrel weighs perhaps 1 kilogram at best. There are few exceptions to the pyramid of biomass scheme. One occurs in aquatic systems where the algae may be both outnumbered and outweighed by the organisms that feed on the algae. The algae can support the greater biomass of the next trophic level only because they can reproduce as fast as they are eaten. In this way, they are never completely consumed. It is interesting to note that this exception to the rule of the pyramid of biomass also is a partial exception to at least 2 of the 3 reasons for the pyramid of biomass. While not all the algae are consumed, a greater proportion of them are, and while not completely digestible, algae are far more nutritious overall than the average woody plant is (most organisms cannot digest wood and extract energy from it).

Biological Magnification

Biological magnification is the tendency of pollutants to become concentrated in successive trophic levels. Often, this is to the detriment of the organisms in which these materials concentrate, since the pollutants are often toxic.

Biomagnification occurs when organisms at the bottom of the food chain concentrate the material above its concentration in the surrounding soil or water. Producers, as we saw earlier, take in inorganic nutrients from their surroundings. Since a lack of these nutrients can limit the growth of the producer, producers will go to great lengths to obtain the nutrients. They will spend considerable energy to pump them into their bodies. They will even take up more than they need immediately and store it, since they can't be "sure" of when the nutrient will be available again (of course, plants don't think about such things, but, as it turns out, those plants, which, for whatever reason, tended to concentrate inorganic nutrients have done better over the years). The problem comes up when a pollutant, such as DDT or mercury, is present in the environment. Chemically, these pollutants resemble essential inorganic nutrients and are brought into the producer's body and stored "by mistake". This is the first step in biomagnification; the pollutant is at a higher concentration inside the producer than it is in the environment.

The second stage of biomagnification occurs when the producer is eaten. Remember from our discussion of a pyramid of biomass that relatively little energy is available from one trophic level to the next. This means that a consumer (of any level) has to consume a lot of biomass from the lower trophic level. If that biomass contains the pollutant, the pollutant will be taken up in large quantities by the consumer. Pollutants that biomagnify have another characteristic. Not only are they taken up by the producers, but they are absorbed and stored in the bodies of the consumers. This often occurs with pollutants soluble in fat such as DDT or PCB's. These materials are digested from the producer and move into the fat of the consumer. If the consumer is caught and eaten, its fat is digested and the pollutant moves to the fat of the new consumer. In this way, the pollutant builds up in the fatty tissues of the consumers. Water-soluble pollutants usually cannot biomagnify in this way because they would dissolve in the bodily fluids of the consumer. Since every organism loses water to the environment, as the water is lost the pollutant would leave as well. Alas, fat simply does not leave the body.

The "best" example of biomagnification comes from DDT. This long-lived pesticide (insecticide) has improved human health in many countries by killing insects such as mosquitoes that spread disease. On the other hand, DDT is effective in part because it does not break down in the environment. It is picked up by organisms in the environment and incorporated into fat. Even here, it does no real damage in many organisms (including humans). In others, however, DDT is deadly or may have more insidious, long-term effects. In birds, for instance, DDT interferes with the deposition of calcium in the shells of the bird's eggs. The eggs laid are very soft and easily broken; birds so afflicted are rarely able to raise young and this causes a decline in their numbers. This was so apparent in the early 1960's that it led the scientist Rachel Carson to postulate a "silent spring" without the sound of bird calls. The banning of DDT, the search for pesticides that would not biomagnify, and the birth of the "modern" environmental movement in the 1960's. Birds such as the bald eagle have made comebacks in response to the banning of DDT in the US. Ironically, many of the pesticides which replaced DDT are more dangerous to humans, and, without DDT, disease (primarily in the tropics) claims more human lives.

HUMAN VS. NATURAL FOOD CHAINS

Human civilization is dependent on agriculture. Only with agriculture can a few people feed the rest of the population; the part of the population freed from raising food can then go on to do all the things we associate with civilization. Agriculture means manipulating the environment to favour plant species that we can eat. In essence, humans manipulate competition, allowing favoured species (crops) to thrive and thwarting species which might otherwise crowd them out (weeds). In essence, with agriculture we are creating a very simple ecosystem. At most, it has only three levels - producers (crops), primary consumers (livestock, humans) and secondary consumers (humans). This means that little energy is lost between tropic levels, since there are fewer trophic levels present.

This is good for humans, but what type of "ecosystem" have we created? Agricultural ecosystems have several problems. First, we create *monocultures*, or fields with only one crop. This is simplest for planting, weeding, and harvesting, but it also packs many similar plants into a small area, creating a situation ideal for disease and insect pests. In natural ecosystems, plants of one species are often scattered. Insects, which often specialize on feeding on a particular plant species, have a hard time finding the scattered plants. Without food, the insect populations are kept in check. In a field of corn however, even the most inept insect can find a new host plant with a jump in any direction. Likewise, disease is more easily spread if the plants are in close proximity. It takes lots of chemicals (pesticides) to keep a monoculture going.

Another problem with human agriculture is that we rely on relatively few plants for food. If the corn and rice crops failed worldwide in the same year, we would be hard-pressed to feed everyone (not that we're doing a great job of it now). Natural ecosystems usually have alternate sources of food available if one fails.

A final problem associated with agroecosystems is the problem of inorganic nutrient recycling. In a natural ecosystem, when a plant dies it fall to the ground and rots, and its inorganic nutrients are returned to the soil from which they were taken. In human agriculture, however, we harvest the crop, truck it away, and flush it down the toilet to be run off in the rivers to the ocean. Aside from the water pollution problems this causes, it should be obvious to you that the nutrients are not returned to the fields. They have to be replaced with chemical fertilizers, and that means mining, transportation, electricity, etc. Also, the chemical fertilizers tend to run off the fields (along with soil disrupted by cultivation) and further pollute the water.

Some solutions are at hand, but they bring on new problems, too. No-till farming uses herbicides to kill plants in a field; the crop is then planted through the dead plants without plowing up the soil. This reduced soil and fertilizer erosion, but the herbicides themselves may damage ecosystems. In many areas, sewage sludge is returned to fields to act as a fertilizer. This reduces the need for chemical fertilizers, but still requires a lot of energy to haul the sludge around. Further, if one is not careful, things such as household chemicals and heavy metals may contaminate the sewage sludge and biomagnify in the crops which we would then eat.

BIOGEOCHEMICAL CYCLES

We have already seen that while energy does not cycle through an ecosystem, chemicals do. The inorganic nutrients cycle through more than the organisms, however, they also enter into the atmosphere, the oceans, and even rocks. Since these *chem*icals cycle through both the *bio*logical and the *geo*logical world, we call the overall cycles biogeochemical cycles. Each chemical has its own unique cycle, but all of the cycles do have some things in common. *Reservoirs* are those parts of the cycle where the chemical is held in large quantities for long periods of time. In *exchange pools*, on the other hand, the chemical is held for only a short time. The length of time a chemical is held in an exchange pool or a reservoir is termed its *residence* time. The oceans are a reservoir for water, while a cloud is an exchange pool. Water may reside in an ocean for thousands of years, but in a cloud for a few days at best. The biotic community includes all living organisms. This community may serve as an exchange pool (although for some chemicals like carbon, bound in a sequoia for a thousand years, it may seem more like a reservoir), and also serve to move chemicals from one stage of the cycle to another. For instance, the trees of the tropical rain forest bring water up from the forest floor to be evapourated into the atmosphere. Likewise, coral endosymbionts take carbon from the water and turn it into limestone rock. The energy for most of the transportation of chemicals from one place to another is provided either by the sun or by the heat released from the mantle and core of the Earth.

While all inorganic nutrients cycle, we will focus on only 4 of the most important cycles - water, carbon (and oxygen), nitrogen, and phosphorous.

The Water Cycle

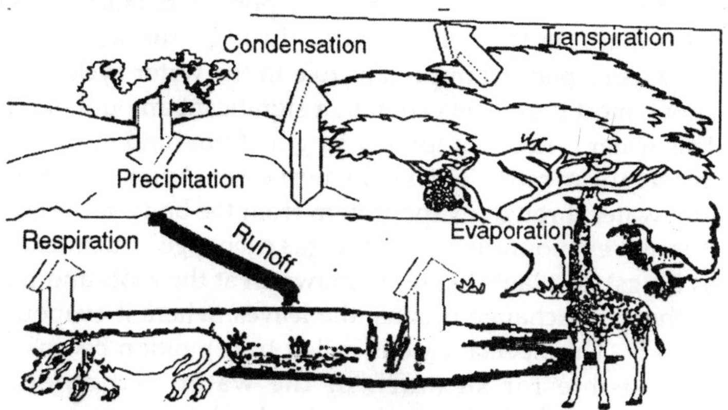

Fig.: Water Cycle

In the water cycle, energy is supplied by the sun, which drives evapouration whether it be from ocean surfaces or from treetops. The sun also provides the energy which drives the weather systems which move the water vapour (clouds) from one place to another (otherwise, it would only rain over the oceans). Precipitation occurs when water condenses from a gaseous state in the atmosphere and falls to earth. Evapouration is the reverse process in which liquid water becomes gaseous. Once water condenses, gravity takes over and the water is pulled to the ground. Gravity continues to operate, either pulling the water underground (groundwater) or across the surface (runoff). In either event, gravity continues to pull water lower and lower until it reaches the oceans (in most cases; the Great Salt Lake, Dead Sea, Caspian Sea, and other such depressions may also serve as the lowest basin into which water can be drawn). Frozen water may be trapped in cooler regions of the Earth (the poles, glaciers on mountaintops, etc.) as snow or ice, and may remain as such for very long periods of time. Lakes, ponds, and wetlands form where water is temporarily trapped.

The oceans are salty because any weathering of minerals that occurs as the water runs to the ocean will add to the mineral content of the water, but water cannot leave the oceans except by evapouration, and evapouration leaves the minerals behind. Thus, rainfall and snowfall are comprised of relatively clean water, with the exception of pollutants (such as acids) picked up as the waster falls through the atmosphere. Organisms play an important role in the water cycle. As you know, most organisms contain a significant amount of water. This water is not held for any length of time and moves out of the organism rather quickly in most cases. Animals and plants lose water through evapouration from the body surfaces, and through evapouration from the gas exchange structures (such as lungs). In plants, water is drawn in at the roots and moves to the gas exchange organs, the leaves, where it evapourates quickly. This special case is called transpiration because it is responsible for so much of the water that enters the atmosphere. In both plants and animals, the breakdown of carbohydrates (sugars) to produce energy (respiration) produces both carbon dioxide and water as waste products. Photosynthesis reverses this reaction, and water and carbon dioxide are combined to form carbohydrates. Now you understand the relevance of the term carbohydrate; it refers to the combination of carbon and water in the sugars we call carbohydrates.

Carbon Cycle

Once you understand the water cycle, the carbon cycle is relatively simple. From a biological perspective, the key events here are the complementary reactions of respiration and photosynthesis. Respiration takes carbohydrates and oxygen and combines them to produce carbon dioxide, water, and energy. Photosynthesis takes carbon dioxide and water and produces carbohydrates and oxygen. The outputs of respiration are the inputs of photosynthesis, and the outputs of photosynthesis are the inputs of respiration.

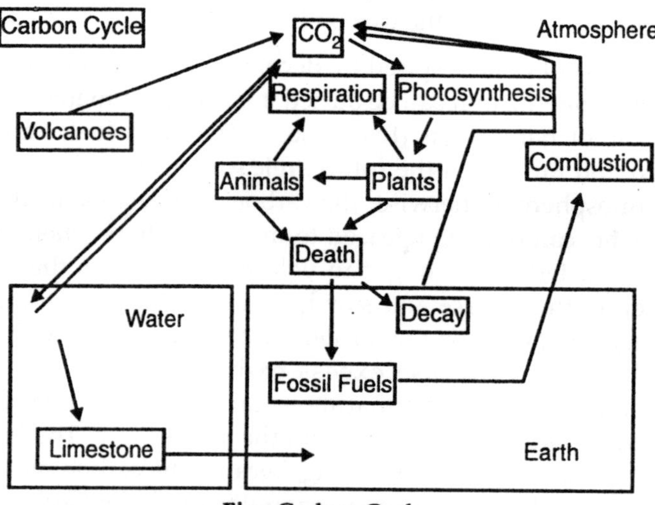

Fig.: Carbon Cycle

The reactions are also complementary in the way they deal with energy. Photosynthesis takes energy from the sun and stores it in the carbon-carbon bonds of carbohydrates; respiration releases that energy. Both plants and animals carry on respiration, but only plants (and other producers) can carry on photosynthesis. The chief reservoirs for carbon dioxide are in the oceans and in rock. Carbon dioxide dissolves readily in water. Once there, it may precipitate (fall out of solution) as a solid rock known as calcium carbonate (limestone). Corals and algae encourage this reaction and build up limestone reefs in the process. On land and in the water, plants take up carbon dioxide and convert it into carbohydrates through photosynthesis. This carbon in the plants now has 3 possible fates. It can be liberated to the atmosphere by the plant through respiration; it can be eaten by an animal, or it can be present in the plant when the plant dies. Animals obtain all their carbon in their food, and, thus, all carbon in biological systems ultimately comes from plants (autotrophs). In the animal, the carbon also has the same 3 possible fates. Carbon from plants or animals that is released to the atmosphere through respiration will either be taken up by a plant in photosynthesis or dissolved in the oceans. When an animal or a plant dies, 2 things can happen to the carbon in it.

It can either be respired by decomposers (and released to the atmosphere), or it can be buried intact and ultimately form coal, oil, or natural gas (fossil fuels). The fossil fuels can be mined and burned in the future; releasing carbon dioxide to the atmosphere. Otherwise, the carbon in limestone or other sediments can only be released to the atmosphere when they are subducted and brought to volcanoes, or when they are pushed to the surface and slowly weathered away. Humans have a great impact on the carbon cycle because when we burn fossil fuels we release excess carbon dioxide into the atmosphere. This means that more carbon dioxide goes into the oceans, and more is present in the atmosphere. The latter condition causes global warming, because the carbon dioxide in the atmosphere allows more energy to reach the Earth from the sun than it allows to escape from the Earth into space.

The Oxygen Cycle

If you look back at the carbon cycle, we have also described the oxygen cycle, since these atoms often are combined. Oxygen is present in the carbon dioxide, in the carbohydrates, in water, and as a molecule of two oxygen atoms. Oxygen is released to the atmosphere by autotrophs during photosynthesis and taken up by both autotrophs and heterotrophs during respiration. In fact, all of the oxygen in the atmosphere is *biogenic*; that is, it was released from water through photosynthesis by autotrophs. It took about 2 billion years for autotrophs (mostly cyanobacteria) to raise the oxygen content of the atmosphere to the 21% that it is today; this opened the door for complex organisms such as multicellular animals, which need a lot of oxygen.

The Nitrogen Cycle:

The nitrogen cycle is one of the most difficult of the cycles to learn, simply because there are so many important forms of nitrogen, and because organisms are responsible for each of the interconversions. Remember that nitrogen is critically important in forming the amino portions of the amino acids which in turn form the proteins of your body. Proteins make up skin and muscle, among other important structural portions of your body, and all enzymes are proteins. Since enzymes carry out almost all of the chemical reactions in your body, it's easy how important nitrogen is. The chief reservoir of nitrogen is the atmosphere, which is about 78% nitrogen.

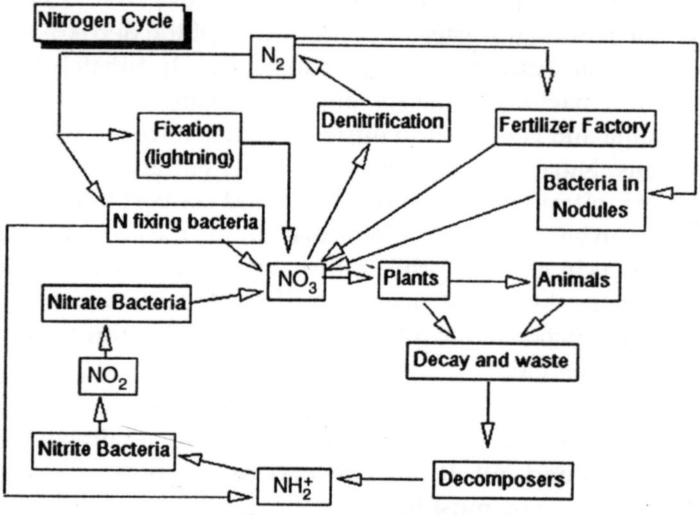

Fig.: Nitrogen Cycle

It is a pretty non-reactive gas; it takes a lot of energy to get nitrogen gas to break up and combine with other things, such as carbon or oxygen. Nitrogen gas can be taken from the atmosphere (fixed) in two basic ways. First, lightning provides enough energy to "burn" the nitrogen and fix it in the form of nitrate, which is a nitrogen with three oxygens attached. This process is duplicated in fertilizer factories to produce nitrogen fertilizers. The other form of nitrogen fixation is by nitrogen fixing bacteria, who use special enzymes instead of the extreme amount of energy found in lightning to fix nitrogen. These nitrogen-fixing bacteria come in three forms: some are free-living in the soil; some form symbiotic, mutualistic associations with the roots of bean plants and other legumes (rhizobial bacteria); and the third form of nitrogen-fixing bacteria are the photosynthetic cyanobacteria (blue-green algae) which are found most commonly in water. All of these fix nitrogen, either in the form of nitrate or in the form of ammonia (nitrogen with 3 hydrogens attached). Most plants can take up nitrate and convert it to amino acids. Animals acquire all of their amino acids when they eat plants (or other animals).

When plants or animals die (or release waste) the nitrogen is returned to the soil. The usual form of nitrogen returned to the soil in animal wastes or in the output of the decomposers, is ammonia. Ammonia is rather toxic, but, fortunately there are nitrite bacteria in the soil and in the water which take up ammonia and convert it to nitrite, which is nitrogen with two oxygens. Nitrite is also somewhat toxic, but another type of bacteria, nitrate bacteria, take nitrite and convert it to nitrate, which can be taken up by plants to continue the cycle. We now have a cycle set up in the soil (or water), but what returns nitrogen to the air? It turns out that there are denitrifying bacteria which take the nitrate and combine the nitrogen back into nitrogen gas.

The nitrogen cycle has some important practical considerations, as anyone who has ever set up a saltwater fish tank has found out. It takes several weeks to set up such a tank, because you must have sufficient numbers of nitrite and nitrate bacteria present to detoxify the ammonia produced by the fish and decomposers in the tank. Otherwise, the ammonia levels in the tank will build up and kill the fish.

This is usually not a problem in freshwater tanks for two reasons. One, the pH in a freshwater tank is at a different level than in a saltwater tank. At the pH of a freshwater tank, ammonia is not as toxic. Second, there are more multicellular plant forms that can grow in freshwater, and these plants remove the ammonia from the water very efficiently. It is hard to get enough plants growing in a saltwater tank to detoxify the water in the same way.

The Phosphorous Cycle

The phosphorous cycle is the simplest of the cycles.

Phosphorous has only one form, phosphate, which is a phosphorous atom with 4 oxygen atoms. This heavy molecule never makes its way into the atmosphere, it is always part of an organism, dissolved in water, or in the form of rock.

When rock with phosphate is exposed to water (especially water with a little acid in it), the rock is weathered out and goes into solution. Autotrophs take this phosphorous up and use it in a variety of ways. It is an important constituent of cell membranes, DNA, RNA, and, of course ATP, which, after all, stands for adenosine tri*phosphate*. Heterotrophs (animals) obtain their phosphorous from the plants they eat, although one type of heterotroph, the fungi, excel at taking up phosphorous and may form mutualistic symbiotic relationships with plant roots.

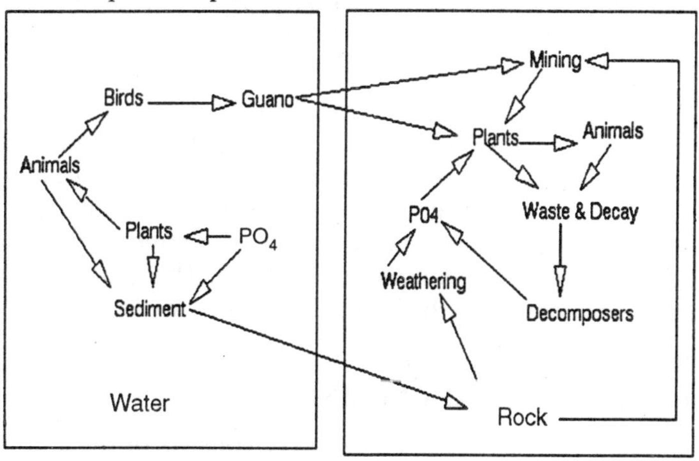

Fig.: Phosphorous Cycle

These relationships are called *mycorrhizae*; the plant gets phosphate from the fungus and gives the fungus sugars in return. Animals, by the way, may also use phosphorous as a component of bones, teeth and shells.

When animals or plants die (or when animals defecate), the phosphate may be returned to the soil or water by the decomposers. There, it can be taken up by another plant and used again. This cycle will occur over and over until at last the phosphorous is lost at the bottom of the deepest parts of the ocean, where it becomes part of the sedimentary rocks forming there. Ultimately, this phosphorous will be released if the rock is brought to the surface and weathered. Two types of animals play a unique role in the phosphorous cycle. Humans often mine rock rich in phosphorous. For instance, in Florida, which was once sea floor, there are extensive phosphate mines.

The phosphate is then used as fertilizer. This mining of phosphate and use of the phosphate as fertilizer greatly accelerates the phosphorous cycle and may cause local overabundance of phosphorous, particularly in coastal regions, at the mouths of rivers, and anyplace where there is a lot of sewage released into the water (the phosphate placed on crops finds its way into our stomachs and from there to our toilets). Local abundance of phosphate can cause overgrowth of algae in the water; the algae can use up all the oxygen in the water and kill other aquatic life.

This is called eutrophication. The other animals that play a unique role in the phosphorous cycle are marine birds. These birds take phosphorous containing fish out of the ocean and return to land, where they defecate. Their guano contains high levels of phosphorous and in this way marine birds return phosphorous from the ocean to the land. The guano is often mined and may form the basis of the economy in some areas.

ENERGY TRANSFORMATIONS AND BIOGEOCHEMICAL CYCLING

The study of ecosystems mainly consists of the study of certain processes that link the living, or biotic, components to the non-living, or abiotic, components. *Energy transformations* and *biogeochemical cycling* are the main processes that comprise the field of ecosystem ecology. As we learned earlier, ecology generally is defined as the interactions of organisms with one another and with the environment in which they occur. We can study ecology at the level of the individual, the population, the community, and the ecosystem.

Studies of *individuals* are concerned mostly about physiology, reproduction, development or behaviour, and studies of *populations* usually focus on the habitat and resource needs of individual species, their group behaviours, population growth, and what limits their abundance or causes extinction. Studies of *communities* examine how populations of many species interact with one another, such as predators and their prey, or competitors that share common needs or resources.

In *ecosystem ecology* we put all of this together and, insofar as we can, we try to understand how the system operates as a whole. This means that, rather than worrying mainly about particular species, we try to focus on major functional aspects of the system. These *functional aspects* include such things as the amount of energy that is produced by photosynthesis, how energy or materials flow along the many steps in a food chain, or what controls the rate of decomposition of materials or the rate at which nutrients are recycled in the system.

COMPONENTS OF AN ECOSYSTEM

A basic understanding of the diversity of plants and animals, and how plants and animals and microbes obtain water, nutrients, and food. We can clarify the parts of an ecosystem by listing them under the headings "abiotic" and "biotic".

Abiotic Components	Biotic Components
Sunlight	Primary producers
Temperature	Herbivores
Precipitation	Carnivores
Water or moisture	Omnivores
Soil or water chemistry (e.g., P, NH_4^+)	Detritivores
etc.	etc.

All of these vary over space/time

By and large, this set of environmental factors is important almost everywhere, in all ecosystems.

A *functional group* is a biological category composed of organisms that perform mostly the same kind of function in the system; all the photosynthetic plants or primary producers form a functional group. Membership in the functional group does not depend very much on who the actual players happen to be, only on what function they perform in the ecosystem.

PROCESSES OF ECOSYSTEMS

This figure with the plants, zebra, lion, and so forth illustrates the two main ideas about how ecosystems function: *ecosystems have energy flows* and *ecosystems cycle materials*. These two processes are linked, but they are not quite the same

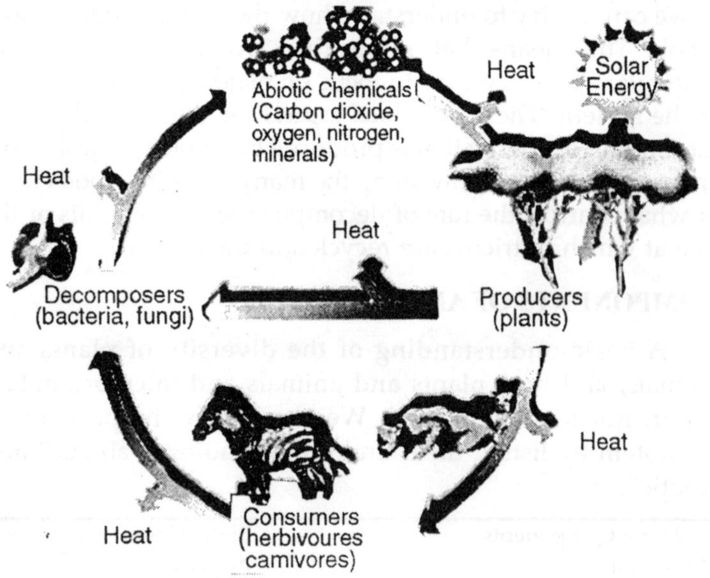

Fig.: Energy flows and material cycles.

Energy enters the biological system as light energy, or photons, is transformed into chemical energy in organic molecules by cellular processes including photosynthesis and respiration, and ultimately is converted to heat energy. This energy is dissipated, meaning it is lost to the system as heat; once it is lost it cannot be recycled. Without the continued input of solar energy, biological systems would quickly shut down. Thus the earth is an *open system* with respect to energy. Elements such as carbon, nitrogen, or phosphorus enter living organisms in a variety of ways. Plants obtain elements from the surrounding atmosphere, water, or soils. Animals may also obtain elements directly from the physical environment, but usually they obtain these mainly as a consequence of consuming other organisms.

These materials are transformed biochemically within the bodies of organisms, but sooner or later, due to excretion or decomposition, they are returned to an inorganic state. Often bacteria complete this process, through the process called decomposition or mineralization During decomposition these materials are not destroyed or lost, so the earth is a *closed system* with respect to elements (with the exception of a meteorite entering the system now and then). The elements are cycled endlessly between their biotic and abiotic states within ecosystems. Those elements whose supply tends to limit biological activity are called *nutrients.*

THE TRANSFORMATION OF ENERGY

The transformations of energy in an ecosystem begin first with the input of energy from the sun. Energy from the sun is captured by the process of photosynthesis. Carbon dioxide is combined with hydrogen (derived from the splitting of water molecules) to produce carbohydrates (CHO). Energy is stored in the high energy bonds of adenosine triphosphate.

The prophet Isaah said "all flesh is grass", earning him the title of first ecologist, because virtually all energy available to organisms originates in plants. Because it is the first step in the production of energy for living things, it is called *primary production Herbivores* obtain their energy by consuming plants or plant products, *carnivores* eat herbivores, and *detritivores* consume the droppings and carcasses of us all. portrays a simple food chain, in which energy from the sun, captured by plant photosynthesis, flows from *trophic level* to trophic level via the *food chain.* A trophic level is composed of organisms that make a living in the same way, that is they are all *primary producers* (plants), *primary consumers* (herbivores) or *secondary consumers* (carnivores). Dead tissue and waste products are produced at all levels. Scavengers, detritivores, and decomposers collectively account for the use of all such "waste" — consumers of carcasses and fallen leaves may be other animals, such as crows and beetles, but ultimately it is the microbes that finish the job of decomposition. Not surprisingly, the amount of primary production varies a great deal from place to place, due to differences in the amount of solar radiation and the availability of nutrients and water.

Energy transfer through the food chain is inefficient. This means that less energy is available at the herbivore level than at the primary producer level, less yet at the carnivore level, and so on. The result is a pyramid of energy, with important implications for understanding the quantity of life that can be supported. Usually when we think of food chains we visualize green plants, herbivores, and so on. These are referred to as *grazer food chains,* because living plants are directly consumed. In many circumstances the principal energy input is not green plants but dead organic matter. These are called *detritus food chains.* Examples include the forest floor or a woodland stream in a forested area, a salt marsh, and most obviously, the ocean floor in very deep areas where all sunlight is extinguished 1000's of meters above. In subsequent lectures we shall return to these important issues concerning energy flow. Finally, although we have been talking about food chains, in reality the organization of biological systems is much more complicated than can be represented by a simple "chain". There are many food links and chains in an ecosystem, and we refer to all of these linkages as a *food web.* Food webs can be very complicated, where it appears that *"everything is connected to everything else"*, and it is important to understand what are the most important linkages in any particular food web.

BIOGEOCHEMISTRY

One obvious way is to study the flow of energy or the cycling of elements. The cycling of elements is controlled in part by organisms, which store or transform elements, and in part by the chemistry and geology of the natural world. The term *Biogeochemistry* is defined as the study of how living systems influence, and are controlled by, the geology and chemistry of the earth. Thus biogeochemistry encompasses many aspects of the abiotic and biotic world that we live in.

There are several main *principles and tools* that biogeochemists use to study earth systems. Most of the major environmental problems that we face in our world toady can be analysed using biogeochemical principles and tools. These problems include global warming, acid rain, environmental pollution, and increasing greenhouse gases. The principles and tools that we use can be broken down into 3 major components: *element ratios, mass balance, and element cycling.*

Element ratios

In biological systems, we refer to important elements as *"conservative"*. These elements are often nutrients. By "conservative" we mean that an organism can change only slightly the amount of these elements in their tissues if they are to remain in good health. It is easiest to think of these conservative elements in relation to other important elements in the organism. In healthy algae the elements C, N, P, and Fe have the following ratio, called the *Redfield ratio* after the oceanographer who discovered it:

C: N: P: Fe = 106: 16: 1: 0.01

Once we know these ratios, we can compare them to the ratios that we measure in a sample of algae to determine if the algae are lacking in one of these limiting nutrients.

Mass Balance

Another important tool that biogeochemists use is a simple mass balance equation to describe the state of a system. The system could be a snake, a tree, a lake, or the entire globe. Using a mass balance approach we can determine whether the system is changing and how fast it is changing. The equation is:

Net Change = Input + Output + Internal Change

In this equation the net change in the system from one time period to another is determined by what the inputs are, what the outputs are, and what the internal change in the system was. The example given in class is of the acidification of a lake, considering the inputs and outputs and internal change of acid in the lake.

Element Cycling

Element cycling describes where and how fast elements move in a system. There are two general classes of systems that we can analyse.

A closed system refers to a system where the inputs and outputs are negligible compared to the internal changes. Examples of such systems would include a bottle, or our entire globe. There are two ways we can describe the cycling of materials within this closed system, either by looking at the rate of movement or at the pathways of movement.

- Rate = number of cycles / time * as rate increases, productivity increases
- Pathways-important because of different reactions that may occur

In an open system there are inputs and outputs as well as the internal cycling. Thus we can describe the rates of movement and the pathways, just as we did for the closed system, but we can also define a new concept called the residence time. The residence time indicates how long on average an element remains within the system before leaving the system.

- Rate
- Pathways
- Residence time, Rt

Rt = total amount of matter / output rate of matter

CONTROLS ON ECOSYSTEM FUNCTION

Now that we have learned something about how ecosystems are put together and how materials and energy flow through ecosystems, we can better address the question of "what controls ecosystem function"? There are two dominant theories of the control of ecosystems. The first, called bottom-up control, states that it is the nutrient supply to the primary producers that ultimately controls how ecosystems function. If the nutrient supply is increased, the resulting increase in production of autotrophs is propagated through the food web and all of the other trophic levels will respond to the increased availability of food (energy and materials will cycle faster). The second theory, called top-down control, states that predation and grazing by higher trophic levels on lower trophic levels ultimately controls ecosystem function. If you have an increase in predators, that increase will result in fewer grazers, and that decrease in grazers will result in turn in more primary producers because fewer of them are being eaten by the grazers. Thus the control of population numbers and overall productivity "cascades" from the top levels of the food chain down to the bottom trophic levels. So, which theory is correct? Well, as is often the case when there is a clear dichotomy to choose from, the answer lies somewhere in the middle.

There is evidence from many ecosystem studies that BOTH controls are operating to some degree, but that NEITHER control is complete. The "top-down" effect is often very strong at trophic levels near to the top predators, but the control weakens as you move further down the food chain. Similarly, the "bottom-up" effect of adding nutrients usually stimulates primary production, but the stimulation of secondary production further up the food chain is less strong or is absent. Thus we find that both of these controls are operating in any system at any time, and we must understand the relative importance of each control in order to help us to predict how an ecosystem will behave or change under different circumstances, such as in the face of a changing climate.

THE GEOGRAPHY OF ECOSYSTEMS

There are many different ecosystems: rain forests and tundra, coral reefs and ponds, grasslands and deserts. Climate differences from place to place largely determine the types of ecosystems. How terrestrial ecosystems appear to us is influenced mainly by the dominant vegetation. The word "biome" is used to describe a major vegetation type such as tropical rain forest, grassland, tundra, etc., extending over a large geographic area. It is never used for aquatic systems, such as ponds or coral reefs. It always refers to a vegetation category that is dominant over a very large geographic scale, and so is somewhat broader than an ecosystem. Every place on earth gets the same total number of hours of sunlight each year, but not the same amount of heat.

The sun's rays strike low latitudes directly but high latitudes obliquely. This uneven distribution of heat sets up not just temperature differences, but global wind and ocean currents that in turn have a great deal to do with where rainfall occurs. Add in the cooling effects of elevation and the effects of land masses on temperature and rainfall, and we get a complicated global pattern of climate.

A schematic view of the earth shows that, complicated though climate may be, many aspects are predictable High solar energy striking near the equator ensures nearly constant high temperatures and high rates of evapouration and plant transpiration. Warm air rises, cools, and sheds its moisture, creating just the conditions for a tropical rain forest. Contrast the stable temperature but varying rainfall of a site in Panama with the relatively constant precipitation but seasonally changing temperature of a site in New York State. Every location has a rainfall- temperature graph that is typical of a broader region.

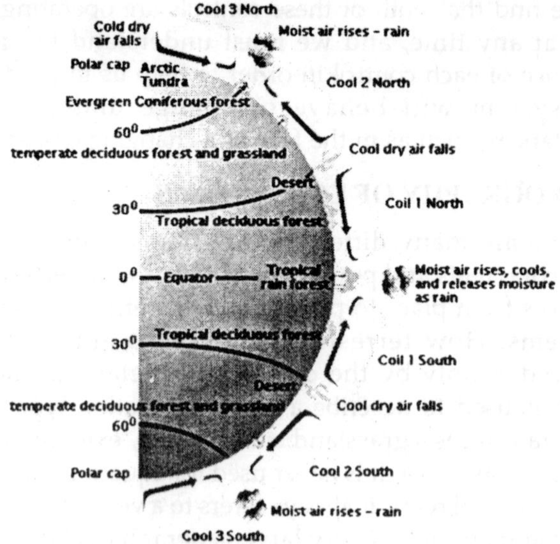

Fig.: Climate patterns affect biome distributions.

We can draw upon plant physiology to know that certain plants are distinctive of certain climates, creating the vegetation appearance that we call biomes. Note how well the distribution of biomes plots on the distribution of climates Note also that some climates are impossible, at least on our planet. High precipitation is not possible at low temperatures — there is not enough solar energy to power the water cycle, and most water is frozen and thus biologically unavailable throughout the year. The high tundra is as much a desert as is the Sahara.

- Ecosystems are made up of abiotic (non-living, environmental) and biotic components, and these basic components are important to nearly all types of ecosystems. Ecosystem Ecology looks at energy transformations and biogeochemical cycling within ecosystems.
- Energy is continually input into an ecosystem in the form of light energy, and some energy is lost with each transfer to a higher trophic level. Nutrients, on the other hand, are recycled within an ecosystem, and their supply normally limits biological activity. So, "energy flows, elements cycle".
- Energy is moved through an ecosystem via a food web, which is made up of interlocking food chains. Energy is first captured by photosynthesis (primary production). The amount of primary production determines the amount of energy available to higher trophic levels.
- The study of how chemical elements cycle through an ecosystem is termed biogeochemistry. A biogeochemical cycle can be expressed as a set of stores (pools) and transfers, and can be studied using the concepts of "stoichiometry", "mass balance", and "residence time".
- Ecosystem function is controlled mainly by two processes, "top-down" and "bottom-up" controls.
- A biome is a major vegetation type extending over a

large area. Biome distributions are determined largely by temperature and precipitation patterns on the Earth's surface.

BIOGEOCHEMISTRY

As the terminology implies, Biogeochemistry is a borderline specialisation involving three basic branches of natural, physical and earth sciences. On one hand living organisms are involved and on the other substances such as nutrients, are made available to the living organisms either through soil (the pedosphere) or through water (the hydrosphere). Of course there are certain organisms which can get their food through air (atmosphere) also. The living organisms, while differing vastly among themselves in size, morphology and physiology, share, all the same a basic feature common to them all: ability to metabolise their requirements in the habitats they live. An individual organism is perishable but life process, sustained by successive generation, is perpetual. Biogeochemistry is all about impact of such living processes on our environment. T

he bulk of living matter is concentrated in the landmass of the world though at the continent-ocean interface and at the bottom of the sea they are present in abundance. The physio-chemical interaction between organisms and the land that sustain them have been a subject of study for over a century now. More than two hundred years ago, The turnover of elements on the surface of the globe in which he postulated on the cyclic exchange of elements among the three kingdoms of nature, namely mineral, vegetable and animal. Lavoisier paid a heavy price for stating so: he was executed after his work was published. The interaction among the above three kingdoms essentially constitute the broad field of Biogeochemistry as we know today.

The original kingdoms have been replaced by reservoirs of different kind and atmosphere is brought into the picture because of gases involved in many biogeochemical processes. Since life is cyclic, all processes where living matter is involved has also got to be cyclic. Hence, in all the reservoirs are linked to each other in both directions:at the inlet and outlet point for each reservoir. The interconnection of all the reservoirs leads to a overall global cycle. Such a cycle is known by various names depending on the user and the sector of interest: biogeochemical cycle, hydrological cycle, geochemical cycle and rock cycle. The emphasis in each cycle varies: in hydrological cycle, one will be interested in transfer of water; in the geochemical cycle, one will be interested in transfer from the crust to other segments and in rock cycle, the interest will be transformation of large volume of crustal materials across various interfaces. Whereas in the biogeochemical cycle the focus is on nutrients essential for the life processes and at the toxic substances determinant to such living things. Thus biogeochemical processes can be defined in three stages:

Stage-1: Elements to be grouped according to the needs:

Essential, such as Carbon (C), Nitrogen (N) and Phosphorous (P)

Minor, such as Sodium (Na), Potassium(K) etc.

Trace, such as Iron (Fe), Manganese (Mn), Copper (Cu) etc.

Toxic, such as Mercury (Hg), Selenium (Se), Cadmium (Cd) (all the above groupings according to the requirements of biological system in the biological world) - the bio part of the topic.

Stage-2: The availability of the above from the earth, generally from the top few metres of the earth (the geocomponent, more specifically the pedosphere the geo part of the topic and now.

Stage-3: The physio-chemical mechanisms that make the needs of the biological systems from the pedospheric layers of the earth in response to micro-environmental functions such as pH (acidity), weathering and release of elements by their breakdown due to chemical interaction of individual minerals with water. This constitutes the chemical part of the topic.

Let us illustrate this linkages using an essential element Carbon as an example. Referring to carbon is present in the atmosphere mainly as carbon dioxide (CO_2) and in small quantities as carbon monoxides (CO). In the hydrosphere, carbon is present as bicarbonate and carbonate the major factor in the hardness of water. In the lithosphere- pedosphere, it is present as calcium carbonate ($CaCO_3$) or some combination thereof. In the biosphere, carbon is present as the building block of the plants and animals in the form of variety of carbon compounds such as carbohydrates, proteins and amino-acids. Carbon enters the biological system either through geological process (weathering of carbon-bearing minerals such as limestone ($CaCO_3$) and /or chemical process directly from the atmosphere (CO_2) by photosynthesis.

When the organism (including plant) dies out, the carbon is released back either directly to the atmosphere as CO_2 or is mineralised as coal, limestone etc. So that it is returned to the lithosphere. Thus the carbon cycle is complete. The oldest carbon-bearing organic world remains known to man is about 3.5 billion years whereas the age of the earth is at present estimated to be about 4.5 billion years. In spite of the complexity of the biological evolution through geological time scale, carbon has been and is being cycled through various reservoirs relatively unaffected and un- disturbed till recent times.

Among the three essential nutrients for life phosphorous (P) is extremely interesting. Unlike carbon and nitrogen, it is predominately derived from the crustal top of the solid earth and mobilised through the biosphere by the transport behaviour in the hydrosphereP ignites in air but glows when left in darkness. Thus it is a carrier of energy in all living systems. The homologous elements P and N has complimentary properties in relation to Hydrogen(H) and Oxygen(O_2)- the other important elements for most living matter on the surface of the earth. P in association with O_2 introduces some structural order in the molecules at cellular levels and thus the P-O association in the form of PO_4 units separates life from ordinary inanimate minerals represented by similar association of Silicon with O_2 (SiO_4)

From a single cell amoeba some 3.5 billion years ago (3500 million), organisms progressively went through complex biological evolution as a function of geological time scale. The link between time and evolution is established through changing environmental parameters such the carbon dioxide (CO_2) levels in atmosphere prevailing at a given time, the O_2 levels, nutrients such as P and minor and trace elements etc. Geochemical evidences through rock records show that about 2.5 billion years ago, the atmospheric CO_2 was at least 100 times lower as that of today and O_2 levels negligible compared to today s environment. have been continuously interacting among each other through such a vast time scale to maintain the cyclic behaviour of nutrients and other elements. Excessive constituents in any one segment was removed over a period of time to other segments so that its biogeochemical cyclic behaviour was not affected. During Gondwana times (about 400 to 500 million years before) when India, Australia and Antarctica were neighbours, excess CO_2 in the atmosphere was neutralised by enhanced photosynthesis and subsequently the biota decayed and gave raise to most of the coal deposits which we are exploiting today.

Thus due to the biogeochemical behaviour, nature was able to adjust itself at its own rate in any parameters. On the other hand in modern times, man introduces any one variable to a segment in very abruptly and in such large dose that nature is unable to adjust to that change. A case in point the emission of green house gas CO_2 by the burning of coal. The essential nutrient element C is transferred from the lithosphere to the atmosphere so fast that the excess CO_2 is unable to be removed equally faster by the hydrosphere leading to enhanced levels of CO_2 in the atmosphere and hence we hear everyday about the global warming On the other hand, let us take a element such as Mercury (Hg) considered to be very toxic to the biological world. The movement of Hg in cyclic fashion through our environment. Hg generally occurs in minerals in association with sulphur in the form of mercury sulphide (HgS) systems.

Even in rocks, it is considered to be trace element since its abundance in the crustal earth is very low. However, during the exploitation of base metals resources, such as that of copper, lead and Zinc, mercury is extracted as a buy-product.

Modern civilisation uses mercury for a variety of products such as chemicals, coatings on seeds etc. Even though it does not react with water in the hydrosphere easily, mercury combines with any organic substances such decayed plant debris etc and immediately becomes toxic to the living biota. The well known minemata decease in Japan was the result of such affinity of mercury for many biotic materials. Being the only metal known to man to be in liquid state at room conditions, it easily vapourises and hence get also into the atmospheric segment Hence due to man s use, what was once locked up safely in inert geological materials in rocks, have been mobilised to the biosphere, hydrosphere and atmosphere.

The story is basically same for a few other toxic metals such as Cadmium though the details may individually vary due to their differing biogeochemical behaviour in our present environment.

It is clear that processes in our environment are better understood by focussing on the interfacing areas. Thus Biogeochemistry is the ideal such frontier area of study and research. It is however a relatively new field and only in the last two decades various researchers across the world have been focussing their attention on this inter- disciplinary subject. Outside India there are a few places where separate institutes exists for this purpose. At about the same time the Ministry of Environment and Forests also approved the setting up of a ENVIS centre in JNU on the same subject under his direction. Though there no other specific institutions in India in this subject, there are individual researchers in many universities working on topics of interest to Biogeochemistry.

Thus, the area of Biogeochemistry is an emerging branch of multi-diciplinary approach involving all major branches of Sciences. From a humble beginning two decades ago, now this subject is well established in the international field and in India is fast catching up as an important tool of research for understanding processes in our environment. Sufficient infra-structure has now developed in India through a variety of individual and institutional efforts so that in the years to come the importance work on Biogeochemistry will be better understood and appreciated.

The involvement of a number of specialists is obvious when ones looks into the above three stages: for understanding the biotic behaviour, one should have a good understanding of ecological principles; the requirements for the ecosystem is provided by basic geochemical processes of weathering helped by microbiological and biochemical reactions in soils and over rocks; farther, from the source to the receptor, chemical and hydrological principles operate as a middleman. Hence within the broad field of Biogeochemistry itself, there are sub-specialisations depending on the background of the researchers- be a ecologist, geochemist or chemical hydrologist. Hence the word Biogeochemistry is truly an integrating field to better understand our present, past and future environmental problems.

Among the more recent sub-specialisations within Biogeochemistry that have emerged in the international research is the application of certain biochemical and molecular biochemical techniques to understand our present and past environment. A case in point is the study about the Alpine-man.

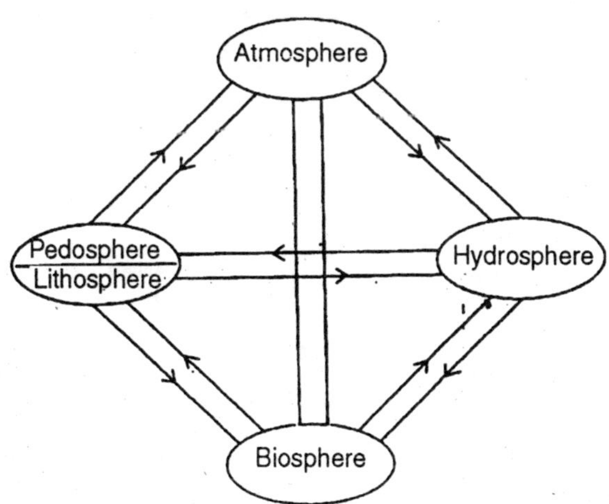

Fig.: Receptors and Donors to other Neighbouring Spheres.

A full human body was found in tact in the frozen alps

and several investigations on the body using most modern biochemical techniques have thrown some light about the habitat of the community some 5000 years ago, then the prevailing eco- system, palaeo-environment, anthropological lineage etc. DNA based study thus offers excellent scope even in environment related matter. In India, a student from our labouratory is working under the supervision of Prof. Laljit Singh, Centre for DNA Finger-printing at Hyderabad to apply DNA techniques in fossils bones of geological and anthropological importance to understand the habitat and palaeo-environment. DNA techniques to ancient biological specimens. Farther, characterisations of amino-acids in ancient sediments have been successfully used to understand palaeo-environment and also stresses acting on such systems. In India, however, this has not been attempted so far.

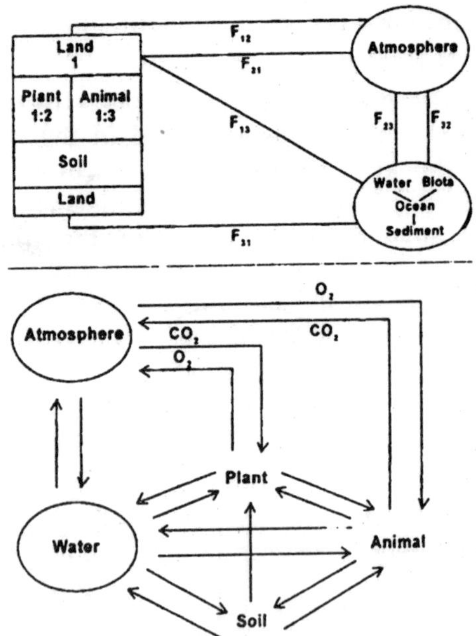

Fig.: Interaction in the Biogeochemical Cycle

Thus, Biogeochemistry can be viewed as a very challenging area of study involving a variety of basic sciences so that it is ideally suited to understand our environment better. Research should not go in the direction of reinventing the wheel but improve and if possible, replace it. Hopefully the 21st century consolidation of the field of Biogeochemistry as an established brach of science.

Enlarged version of specific component. whenevr possible. Important components in each segment and direction of transfer F means flux that is concentration for a given parameter; for ex. the volume of water/air transferred in that direction per unit time. This, F has unit of wt/vol/time. F is likely to have volumes since the time involved is relatively shorter for some of the sectors. For eg. some of the variables transferred involving soil and plants/animals and plants and atmosphere etc. may have a time scale of a few seconds only. The carbon transfer, the following fluxes have been incorporated: 1 to 34 point all have same magnitude, time scale and importance. Between river and young soils sediments, the transfer 26 and 27 represent deposition and erosion processes.

Faster compared to deposition or accumulation so that net transfer is loss of carbon from soil (Pedosphere) to river (hydrosphere). Similarly, due to deforestation, evapouration may be higher than the net river volume.

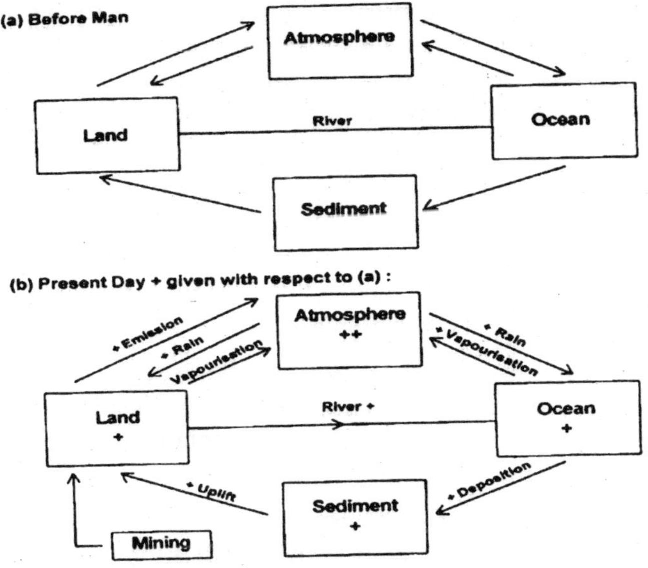

Fig.: Hg in Biogeochemical Cycle

Hg cycle is shwn in tow stages: pre-man and present day.

this concept is adapted from Garrels et al., 1976. The main differences is indicated in teh Hg - mining in the present day cause additional flux - vaporisation - from land to atmosphere. ++ or + singns indicate higher gains in the respectinve sectors i the present day cycle relative to pre-man cycle. As a consequence there is also poportional increase in the flxes in the present day cycle between the concerned sectors. Garels et al. (1976) estimate that Hg will be the first metal that is likely to be lost out entiely ot of the solid spheres (litho - and pedospheres) to the fluid (hydro- and atmo-) spheres in foreseabe future.

MERCURY IN THE ENVIRONMENT

Mercury is a silvery, liquid metal at room temperature and is often referred to as one of the "heavy metals." Like water, mercury can evapourate and become airborne. Because it is an element, mercury does not break down into less toxic substances. Once mercury escapes to the environment, it circulates in and out of the atmosphere until it ends up in the bottoms of lakes and oceans. Mercury can be found as the elemental metal or in a wide variety of organic and inorganic compounds. Depending on its chemical form, mercury may travel long distances before it falls to earth with precipitation or dust.

Bacteria and chemical reactions in lakes and wetlands can change the mercury into a much more toxic form known as methylmercury. Fish become contaminated with methylmercury by eating food (plankton and smaller fish), which has absorbed methylmercury.

As long as the fish continue to be exposed to mercury, mercury continually builds up in fish's bodies. Fish that eat other fish become even more highly contaminated. Thus, the largest tend to be the most contaminated.

When people eat the contaminated fish, the methylmercury can remain in their bodies for a long time. If they eat fish containing methylmercury faster than their bodies can get discharge it, the methylmercury accumulates in their bodies and can be toxic. Many states have fish consumption advisories to inform people about how many meals of fish they can safely eat over a period of time.

Mercury's Environmental Effects

Fish are the main source of food for many birds and other animals, and mercury can seriously damage the health of these species. Loons, eagles, panthers, otters, mink, kingfishers and ospreys naturally eat large quantities of fish. Because these predators rely on speed and coordination to obtain food, mercury may be particularly hazardous to these animals.

Recent research in Minnesota indicates that the following environmental effects are occurring:

- Loons are accumulating so much mercury that it may be affecting their ability to reproduce;
- Elevated levels of mercury have been found in mink and otters;
- Walleye reproduction may be impaired by the fish's exposure to mercury.

Similar effects are being documented for other fish and fish-eating species around the United States and Canada. Has there always been mercury contamination, or is this a recent problem? This is a difficult question to answer, in part because of a lack of adequately preserved fish specimens of preindustrial age to compare against contemporary samples. However, several lines of evidence from recent studies on Wisconsin lakes suggest that increased emissions to the atmosphere, and subsequent higher deposition rates to lakes, likely translate into higher mercury levels in fish.

The Mercury Cycle and Bioaccumulation

There is a constant biogeochemical cycle of mercury. This cycle includes:

- Release of elemental mercury as a gas from the rocks and waters (degassing);
- Long-range transport of the gases in the atmosphere;

- Wet and dry deposition upon land and surface water;
- Absorption onto sediment particles;
- Bioaccumulation (or biomagnification) in terrestrial and aquatic food chains.

Bioaccumulation means an increase in the concentration of a chemical in an organism over time, compared to the chemical's concentration in the environment. Bioaccumulation can be a normal and essential process for the growth of any species, but the accumulation of unnecessary chemicals or toxins, or even the overaccumulation of essential substances can be detrimental. All animals, including humans, daily bioaccumulate many vital nutrients, such as vitamins A, D, and K, trace minerals, essential fats and amino acids, but unfortunately, they can also accumulate many unnecessary substances, such as lead or mercury. What concerns toxicologists is the bioaccumulation of necessary substances to levels in the body that can cause harm. With substances such as lead or mercury, any accumulation at all can be harmful. Compounds accumulate in living things any time they are taken up and stored faster than they are broken down (metabolized) or excreted.

Understanding the dynamic process of bioaccumulation is important in protecting humans and other organisms from the adverse effects of chemical exposure, and it has become a critical consideration in the regulation of chemicals.

Bioaccumulation varies among individual organisms as well as among species. Large, fat, long-lived individuals or species with low rates of metabolism or excretion of a chemical will bioaccumulate more than small, thin, short-lived organisms. Thus, an old lake trout may bioaccumulate much more than a young bluegill in the same lake.

Above is a schematic drawing of mercury cycling in an aquatic ecosystem. With the exception of isolated cases of known point sources, the source of most mercury to most aquatic ecosystems is deposition from the atmosphere, primarily associated with rainfall.

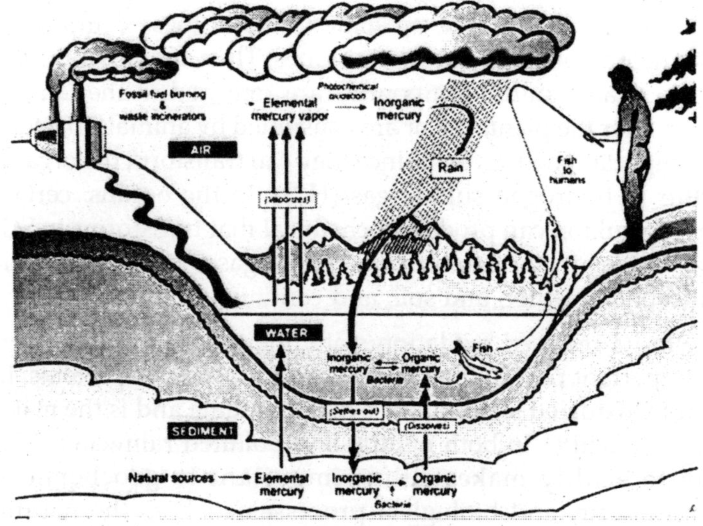

Fig: Mercury Cycle

In the aquatic environment, mercury can be:

- Dissolved or suspended in the water
- Trapped in the sediments
- Ingested by living things (biota)

Methylmercury is the form of mercury most available and most toxic to biota (including zooplankton, insects, fish, and humans). This form of mercury is easily taken up by biota and bioaccumulates in their tissues. Unlike many other fish contaminants, such as PCBs, dioxin, and DDT, mercury does not concentrate in the fat, but in the muscle tissue. Thus, there is no simple way to remove mercury-contaminated portions from fish that is to be eaten.

LIFE AND BIOGEOCHEMICAL CYCLES

THE SULFUR CYCLE

Another example of a major biogeochemical cycle of significance to climate and life is the sulfur cycle. Living things require certain safe, low levels of this nutrient. The sulfur cycle can be thought of as beginning with the gas sulfur dioxide (SO_2) or the particles of sulfate (SO_4=) compounds in the air.

These compounds either fall out or are rained out of the atmosphere. Plants take up some forms of these compounds and incorporate them into their tissues. Then, as with nitrogen, these organic sulfur compounds are returned to the land or water after the plants die or are consumed by animals. Bacteria are important here as well since they can transform the organic sulfur to hydrogen sulfide gas (H_2S). In the oceans, certain phytoplankton can produce a chemical that transforms to SO_2 that resides in the atmosphere. These gases can re-enter the atmosphere, water, and soil, and continue the cycle.

In its reduced oxidation state, the nutrient sulfur plays an important part in the structure and function of proteins. In its fully oxidized state, sulfur exists as sulfate and is the major cause of acidity in both natural and polluted rainwater. This link to acidity makes sulfur important to geochemical, atmospheric, and biological processes such as the natural weathering of rocks, acid precipitation, and rates of denitrification. Sulfur is also one of the main elemental cycles most heavily perturbed by human activity. Estimates suggest that emissions of sulfur to the atmosphere from human activity are at least equal or probably larger in magnitude than those from natural processes. Like nitrogen, sulfur can exist in many forms: as gases or sulfuric acid particles. Sulfuric acid particles contribute to the polluting smog that engulfs some industrial centres and cities where many sulfur containing fuels are burned. Such particles floating in air (known as sulfate aerosols) can cause respiratory diseases or cool the climate by reflecting some extra sunlight to space.

The lifetime of most sulfur compounds in the air is relatively short (e.g. days). Superimposed on these fast cycles of sulfur are the extremely slow sedimentary-cycle processes or erosion, sedimentation, and uplift of rocks containing sulfur. In addition, sulfur compounds from volcanoes are intermittently injected into the atmosphere, and a continual stream of these compounds is produced from industrial activities. These compounds mix with water vapour and form sulfuric acid smog. In addition to contributing to acid rain, the sulfuric acid droplets of smog form a haze layer that reflects solar radiation and can cause a cooling of the earth's surface. While many questions remain concerning specifics, the sulfur cycle in general, and acid rain and smog issues in particular are becoming major physical, biological, and social problems.

BIOGEOCHEMICAL CYCLES

The Earth is a closed system for matter, except for small amounts of cosmic debris that enter the Earth's atmosphere. This means that all the elements needed for the structure and chemical processes of life come from the elements that were present in the Earth's crust when it was formed billions of years ago. This matter, the building blocks of life, continually cycle through Earth's systems, the atmosphere, hydrosphere, biosphere, and lithosphere, on time scales that range from a few days to millions of years. These cycles are called biogeochemical cycles, because they include a variety of biological, geological, and chemical processes. Many elements cycle through ecosystems, organisms, air, water, and soil. Many of these are trace elements. Other elements, including carbon, nitrogen, oxygen, hydrogen, sulfur, and phosphorus are critical components of all biological life. Together, oxygen and carbon account for 80 percent of the weight of human beings. Because these elements are key components of life, they must be available for biological processes.

Carbon, however, is relatively rare in the Earth's crust, and nitrogen, though abundant in the atmosphere, is in a form that is not useable by living organisms.The biogeochemical cycles transport and store these important elements so that they can be used by living organisms. Each cycle takes many different pathways and has various reservoirs, or storage places, where elements may reside for short or long periods of time. Each of the chemical, biological, and geological processes varies in their rates of cycling. Some molecules may cycle very quickly depending on the pathway. Carbon atoms in deep ocean sediments may take hundreds to millions of years to cycle completely through the system. An average water molecule resides in the atmosphere for about ten days, although it may be transported many miles before it falls back to the Earth as rain.

How fast substances cycle depends on its chemical reactivity and whether or not it can be found in a gaseous state. A gaseous phase allows molecules to be transported quickly. Phosphorous has no gaseous phase and is relatively uncreative, so it moves very slowly through its cycle. Phosphorus is stored in large amounts in sediment in the oceans or in the Earth's crust and is recycled back to the surface only over very long periods of time through upwelling of ocean waters or weathering of rocks.

Biogeochemical cycles are subject to disturbance by human activities. Humans accelerate natural biogeochemical cycles when elements are extracted from their reservoirs, or sources, and deposited back into the environment (sinks). Humans have significantly altered the carbon cycle by extracting and combusting billions of tons of hydrocarbons in fossil that were buried deep in the Earth's crust, in addition to clearing vegetation that stores carbon. Global release of carbon through human activities has increased from 1 billion tons per year in 1940 to 6.5 billion tons per year in 2000. About half of this extra carbon is taken up by plants and the oceans, while the other half remains in the atmosphere.

In addition to carbon cycle, humans have altered the nitrogen and phosphorus cycles by adding these elements to croplands as fertilizers, which has contributed to over-fertilization of aquatic ecosystems when excess amounts are carried by runoff into local waterways.

Researchers are trying to understand all of the various pathways and flows of each of the biogeochemical cycles in order to understand how human activities affect these cycles. While many important processes have been understood for more than century, there are many phenomena that scientists are just beginning to investigate. Satellite technology, among other tools, has revealed new information about interactions between the oceans and atmosphere that contribute to knowledge about the carbon cycle, but there remain many unanswered questions.

HUMAN MODIFICATIONS OF CLIMATE SERVICES

Human activities are significantly perturbing all of these biogeochemical cycles as well as other earth system processes, both directly through industrial processes and indirectly through changing distributions and abundance of life. The atmosphere is of particular importance to the perturbations due to its crucial role in mediating all energy that enters and leaves earth. Overall, the atmosphere is the component that controls the dominant energy flow in the earth's climate system, and solar radiation from the sun provides the energy to make the weather machine work. Embedded in this process are the biogeochemical cycles we have described that operate on a variety of time and space scales and help to regulate flows of energy and materials throughout the earth system Yet, while we understand much about the functioning of separate parts of this system, there is still a great deal to be discovered about the feedbacks and linkages that allow these interconnected parts to function as a whole and, in turn, how they will respond to human modification.

Human Disturbance

Life influences the amount of CO_2 in the atmosphere through photosynthesis, respiration, and oceanic absorption. As ecosystems are altered, the balance of these processes will be altered. Human activities are upsetting this balance and increasing CO_2 in the atmosphere through the burning of fossil fuels and clearing of forests.A significant increase in CO_2 could have dramatic consequences. Mathematical models of the climate suggest that when CO_2 doubles (sometime in the middle of the next century should population, economic and technology trends continue as typically projected), the world will warm up somewhere between 1 and 5 degrees Celsius by 2100 A.D. unless other factors counteract or amplify the CO_2 -induced change Even the lower end of that range is a projected warming at the rate of one degree per hundred years, a factor of ten faster than the one degree per thousand that has been the typical average rate of natural sustained global temperature change from the end of the ice ages to warmer interglacial times. Should the higher end of the 1 to 5 degree Celsius range occur, then we could see rates of climate change some fifty times faster than sustained, natural average conditions. Climate largely determines the types of ecosystems that occupy an area. Global climate change at such a rapid rate would force many species to shift their ranges in an attempt to keep up with changing climatic conditions, as occurred during the ice age - interglacial transition ten to fifteen thousand years ago.

Migrations of species such as slow growing trees with large seeds would have to occur much faster than they did in the past to keep up with rapidly shifting climates. Other species could move more easily, raising the likelihood that communities of species could be disassembled. Estimating the rates of global warming in the next century, however, are very controversial because of the uncertainties involved with multiple interacting feedback mechanisms Humanity can control climate in ways other than changing greenhouse gas concentrations. Consider the amount of moisture released to the atmosphere through transpiration in the tropical rainforests.

The dense vegetation in areas such as the Amazon basin typically recycles the precipitation that falls on it many times over, helping to form heavy cloud cover in the region. The clouds, in turn, reflect sunlight and produce more rain, directly influencing regional climate as well as indirectly perturbing global climate through altering large-scale circulation patterns over the tropics. As humanity deforests regions like the Amazon, not only is CO_2 released into the atmosphere, but changes in the hydrologic cycle will almost certainly affect regional climate and possibly even global climatic patterns. In deforested areas of northeastern Brazil, the cutting of the tropical forests has led to desertification, changing both surface reflectivity and the rate of transpiration. This change in ecosystem character can lead to a destabilizing positive feedback, which may cause an even further reduction in precipitation.

In a recent study on the possible climatic impacts of tropical deforestation, researchers suggest that conversion of forest into crop land or pastures would cause significant changes in the local microclimate Expected changes include reduction in soil moisture, larger diurnal fluctuation of surface temperature and humidity deficit, and increased surface runoff during the rainy season and decreased runoff during the dry season. Results from general circulation model simulations of large-scale deforestation and conversion to grassy vegetation in the Amazon basin indicate an increase in surface temperature, decrease in evapotranspiration, and significant reduction in precipitation.

Depending on the scale of the disturbed areas, local climate changes can lead to regional climate changes which, in turn, may cause alterations in the global climate through atmospheric connections between tropical circulation and large-scale circulation patterns outside of the tropics. The effect on the ecological systems through changes in the hydrological cycle, an increase in the dry season, and the disruption of plant-animal interactions may make it difficult for the rainforests to re-establish themselves if they are destroyed. Climate change aside, the implications of this scenario for the conservation of biodiversity are serious.

The provision of fresh water and regulation of its flows through precipitation, evapouration, transpiration, and run off is mediated by all ecosystems. Forests and other vegetation types are critical components of this ecosystem service providing free flood and drought relief among other things. The loss of these services, through landuse change, can exacerbate disasters like spring floods in the Midwest and Southeast resulting from large expanses of land cleared for agriculture as well as the drainage of wetlands and swamps which otherwise might have acted as reservoirs for holding excess water or filtering toxic wastes.

Valuing Carbon

There is already a historic background on the evaluation of carbon that includes climate change policies such as the introduction of carbon taxes that reduce greenhouse gas emissions through increasing prices of carbon-based fuels proportional to the amount of carbon they emit Another mechanism for reducing greenhouse gas emissions is through an international tradable emissions permit system intended to limit emissions of certain pollutants. These policies and others constitute ways of balancing the economic costs of emissions with some assumed benefit of averting the loss of ecosystem services (called "climate damage").

Any comprehensive attempt to evaluate the societal value of climate change should include such things as loss of species diversity, loss of coastline from increasing sea level, environmentally displaced persons, and agricultural losses. first estimated the climate damage at 1% reduction in GNP based on market sector losses for a central estimate of climate change. This was criticized as too narrow a view of climate as a type of public good since it reflected neither non-market values (e.g. species loss) nor climate "surprise" scenarios conducted a survey of conventional economists, environmental economists, atmospheric scientists, and ecologists.

Their estimates of loss of gross world product (GWP) resulting from a 3 degree Celsius warming by 2090 varied between a loss of 0 and 21% of GNP with a mean of 1.9%). For a 6 degree Celsius warming scenario, the respondents predicted a loss of the world economy ranging from 0.8 to 62 % with a mean estimate of 5.5%. A striking difference was noted between respondents from different academic disciplines, with natural scientists' estimates of economic impact 20 to 30 times higher than conventional economists'. Even a two percent loss of GWP in 1995, however, represents climate damage of hundreds of billions of dollars annually.

While it is impossible to estimate credibly a numerical value on all of the ecosystem services provided through the maintenance of the carbon cycle at present state, it may be useful to look at land use change and loss of biomass, mostly through deforestation, as a source of atmospheric CO_2. In a very simplistic and preliminary evaluation, we can use the rates of net deforestation to calculate a value for carbon. Global loss of above-ground biomass from deforestation in the tropics is approximately 1-3 gigatons/year over the past 10 years.

This amounts to between 2-5 gigatons of carbon in carbon dioxide released into the atmosphere each year from deforestation and forest degradation (this does not include the 6 gigaton carbon emissions from the burning of fossil fuels). Much of the carbon from biosphere emissions is taken up immediately by vegetation, however, leaving approximately 1-2.5 gigatons of net carbon added to the atmosphere each year. We can apply the concept of carbon taxation for emissions to an ecosystem service valuation of retaining the carbon in the forests.

Using a range of carbon taxes from typical macroeconomic models between $1 per ton and $100 per ton of carbon, the net value of carbon lost each year amounts to between $1 and $250 billion/year. However, use of optimizing economic models to estimate climate damage is highly unsatisfying since these studies use very limited and often ad hoc assumptions that both over and underestimate the likely damages to various market and non-market sectors.

Methods of Valuation

The need for alternative methods of evaluation of these climate-related ecosystem services is quite clear when examining preliminary public opinion responses of global warming. In a controversial method called contingent valuation respondents are surveyed to determine how much they would be willing to pay to prevent a given global climate change scenario from happening or accept if so much change were to be allowed. The difficulties with this type of valuing of environmental goods and processes are immense, especially since much of the evaluation is subjective. Public opinion depends, in part, upon people's exposure to the issues and the level of education and information on these issues they have received.

In a Southern California study, the contingent valuation technique was applied to the determine the influence of potential changes in temperature and precipitation resulting from global warming on respondents willingness to pay. Factorial survey methods were used to present a variety of hypothetical climate scenarios to a sample of 600 Southern California residents. Respondents were provided with a baseline microclimate for the region before future climate scenarios were evaluated. For residents living in coastal communities, the baseline climate over the past 10 years was described as having a summer average high temperature of 75 degrees Fahrenheit, with daily highs ranging between 70 and 80 degrees, and an average of 13 inches per year of rain. One possible future scenario over the next 10 years included a summer average high temperature of 100 degrees Fahrenheit, with daily highs ranging from 80 to 120 degrees (the latter typical of Death Valley, California), and an average of 20 inches per year of rain.

With these and other scenarios, predicted probabilities were determined from the respondents willingness to pay for the abatement of different mean high temperatures. In this scenario, respondents were willing to pay an average $140 to offset a mean high temperature of 100 degrees, while a mean high temperature of 80 degrees was worth approximately $100 This represents a 40% increment in willingness to pay for a 20 degree rise in temperature and other scenario characteristics. residents reached a plateau in their willingness to pay to prevent 120 degree Fahrenheit mean temperatures as compared to 110 degree Fahrenheit mean temperatures.

However, the actual damages to the L.A. basin residents of mean high temperatures of 110 or more degrees Fahrenheit (which would imply occasional extreme heat waves similar in temperature to Death Valley mean highs) would be orders of magnitude more costly, we believe, than a 100 degree Fahrenheit mean high temperature, as such extreme heat would decimate most existing vegetation and threaten the lives of tens of thousands of elderly and other persons vulnerable to heat stroke. For just such reasons, Berk and Shulman strongly caution against taking the dollar values from the survey literally or using them in cost-benefit analyses as they confound several source of value including stewardship and altruism. In addition, some of the climate increases were well above the range of current scientific estimates of greenhouse warming.

The survey was not done in conjunction with atmospheric scientists and climatologists who could provide more realistic climate scenarios or ecologists, public health officials, or others who could help the respondents realise what such warming might mean for trees, birds, or people. Contingent valuation of the hypothetical good is possible when people believe the survey scenario. We present this type of evaluation study to highlight how difficult it is to find acceptable methods to place values on the climatic components of ecosystem services. In this survey case, the background of the respondents as well as their (limited) prior knowledge of the impacts of greenhouse warming played a large role in the survey outcomes. At the same time, however, contingent valuation points out that people are willing to pay to preserve ecosystem services as well as the tremendous need for education to help citizens more realistically value climate and other environmental

The ongoing disturbances of the atmosphere that affect the biogeochemical and physical processes that determine the climate may influence human and natural systems in profound ways. We have attempted to outline a few of the major ecosystem services that are associated with climate and the atmosphere, as well as introduce the challenging task of quantifying, and ultimately monetizing, these services. Current monetized estimates of climate damage by the middle of the 21st century from typical climate change scenarios range from slight economic benefit to a trillion or more dollars lost annually, with most macroeconomic assessments assuming a one to two percent annual loss to GWP from climate change. Moreover, the interacting processes and biogeochemical cycles occurring across a wide spectrum of scales lead to synergistic effects that are not usually considered and sometimes not even known when we attempt to disaggregate and value ecosystem services. The deforestation of the Amazon basin is one example of interacting scales where land use change affecting local and regional climate may also produce a net global residual. Even if the mosaic of regional effects average themselves out globally, there could be residual effects arising from heterogeneous forcing of the climate in areas outside of the tropics (i.e. regional high concentrations of sulfate aerosols or tropospheric ozone).

Ecosystems both mediate and respond to the climate system through a variety of physical, biological, and chemical feedback cycles. The uncertainty of resulting synergisms and potential global effects, as exemplified in the Amazon basin, points to the important challenge of defining and understanding the processes that link species and ecosystems with climate. With increasing knowledge, we can better anticipate ecological responses under changing climate scenarios. Meanwhile, humanity continues to perform this potentially trillion dollar unnatural experiment on "Labouratory Earth".

AIR, CLIMATE AND WEATHER

The Earth's atmosphere is a blanket of gases approximately 350km (218 miles) thick. It is a large and complex system that interacts with the Sun, the land, and the oceans in order to produce both the Earth's weather and climate.

The Earth's atmosphere has four distinct layers

- Thermosphere
- Mesosphere
- Stratosphere
- Troposphere

Thermosphere

The layer of atmosphere most distant from the Earth is the thermosphere, which begins approximately 80km in altitude. It is also the hottest layer; "thermo" being Greek for heat. The temperatures in the thermosphere increase with altitude due to the absorption of intense solar radiation by the limited amount of remaining molecular oxygen. The source of this heat is through bombardment of solar particles carried on the solar wind that do not reach deeper into the atmosphere.

Mesosphere

The mesosphere extends from 50 to 80km in altitude with very sparse atmosphere, accounting for only about 0.1 percent of the mass of the atmosphere as a whole. Temperatures decline within the mesosphere as altitude rises, containing the coldest temperatures within the Earth's atmosphere. At its upper boundary, the mesopause, average temperatures are near -110° C in the summer and -60° C in the winter.

Stratosphere

The stratosphere extends from approximately 10-12 km to around 50 km above the Earth's surface. The air temperature remains relatively constant up to an altitude of 25 km, then increases gradually having a stabilizing effect on atmospheric conditions. The stratosphere contains nearly 90 percent of the atmospheric ozone, which plays a major role in regulating temperatures as solar energy is converted to kinetic energy when the ozone molecules absorb ultraviolet radiation, resulting in the heating of the stratosphere.

Troposphere

The troposphere is the layer closest to the Earth's surface, containing more than 80 percent of total atmospheric mass - composed of 78 percent nitrogen, 21 percent oxygen, other trace gases, water droplets, dust, and other particles. The troposphere is where most weather occurs; the circulation of air in intensive vertical movements results in the formation of clouds, while horizontal movements results in wind. Both temperature and water vapour content decrease rapidly as altitude increases within the troposphere. Nearly 99 percent of atmospheric water vapour is contained within this layer, which plays a major role in regulating air temperature as it absorbs solar energy and thermal radiation from the planet's surface.

Scientists believe that the Earth's climate is changing and is, in fact, heating up. However, there are considerable differences among the views with regard to the rate of change, the impact on our environment, and what can or should be done about it. Current environmental concerns are increasingly focusing on the particles and gases released into the atmosphere as a result of human activity. Frequently, unusual weather events, such as Hurricane Katrina in 2005, are being linked to global warming. But "weather" and "climate" are two very different things, and it is important to understand the difference, as well as the relation, between the two.

In the troposphere, air rises as it is heated by the sun, falls earthward as it cools, and intermixes with evapourated water from the planet's bodies of water to form clouds and precipitation. The uneven heating of the Earth's surface by the sun, along with Earth's rotation, creates rising (convection), falling (advection), and horizontal air movements (winds). The result of these processes occurring in the form of rain, snow, heat or freezing cold, at a particular place and time, is called weather. The longer term trends in patterns of temperature, rainfall, and other weather indicators over time, usually in blocks of 30 years that can affect the entire Earth, is called climate.

Land and Land Use

Land is the solid part of the Earth's surface, also known as the lithosphere. The surface of the Earth is shaped by a combination of physical processes, including earthquakes and volcanoes, shifts of rocks and sediments, and flows of river and ice. Human activities also shape the land in many ways, including expanding agriculture and population, excavating mineral and forest resources, and changing the flow of rivers with dams and channels. Land cover is the physical and biological material found on the land surface – which can be either vegetation or man-made structures. Land cover differs from land use, although the way land is used is often either a cause or an effect of its cover type. Land use describes the various ways in which human beings make use of and manage the land and its resources. The way land is used and managed can alter land cover, and can also have an impact on surrounding habitats, the flow of surface waters, and on biogeochemical cycles.

Land cover and land use issues, therefore, can play a significant role in environmental change, including climate change. When land cover is altered, it can result in the destruction of natural carbon sinks that absorb atmospheric carbon dioxide, a heat-trapping gas that contributes to climate change. Deforestation, is a major contributor to observed increases of carbon dioxide in the atmosphere.

There are often competing interests and tradeoffs related to land use that must be taken into consideration. Land use planners and decision makers often struggle with how to meet human needs for food and the utilization of our natural resources while also minimizing impacts and preserving important habitats. These choices are typically made by those who own or control the land; although the choices can also be limited by the physical and biological characteristics of the land, including the climate, soils, and geography.

Land use choices are also limited by institutional factors and, in the United States, can be governed by a mix of local, county, state, and federal laws. Although most of the land. is privately owned, the federal government is the single largest owner of land.

Analysis of land use and land cover has been greatly facilitated in the last few decades with the availability of new technologies, including remote sensing. The development of geographic information systems (GIS) has made it possible to combine physical characteristics of the land with information about population, economic data, and other information for further analysis

Water

"Water, water, everywhere, nor any drop to drink."

More than 70 percent of the Earth's surface is covered by

water. When the Apollo space missions beamed back pictures of the Earth, public attention was caught by the image of a "Blue Planet," because of the vast expanses of ocean covering the globe. Yet despite this seemingly inexhaustible supply (approximately 370 billion billion gallons) of water on earth, access to clean drinking water remains a major challenge for a large percentage of the world's population. Almost 97 percent of the Earth's water is in the oceans and is too salty to drink. Less than 1 percent of the water on the earth is fresh water (the remaining two to three percent is frozen fresh water contained in glaciers and ice caps).

Water is indispensable for life, not only for drinking water, but also for raising crops and animals for food. Fishing provides a major source of protein for much of the world's population. Throughout history, cities and villages have grown up near sources of water, for drinking, sustenance, and transportation.

In this century, population growth has increased in areas where fresh water is relatively scarce. Water in these areas is a valuable commodity and, in some cases, great technological feats have been undertaken to provide a reliable source of water. In some areas, water quality is a serious concern and waterborne diseases remain a significant source of mortality in many developing countries. The United States has made considerable progress over the last few decades in reducing direct discharges of pollutants into lakes, rivers, and streams. However, indirect sources of pollutants, including both nutrients and toxic substances that are carried by flowing surface waters into larger bodies of water, have proven more difficult to control.

Ecosystems

An ecosystem is generally defined as a community of organisms living in a particular environment and the physical elements in that environment with which they interact. But where does one particular ecosystem end and another begin? An ecosystem can be as small as a field or as large as the ocean, depending on the scale that the researcher is examining. The borders of an ecosystem may be clear, such as a pond; other borders may be less easy to define, such as grassland that gradually changes into brush.

Just as there is an immense diversity of individual species on the planet, so is there a rich diversity of ecosystems, from the icy arctic zones to tropical forests lush with plants and animals. Even the depths of the oceans, once thought to be barren, are now known to be teeming with living microorganisms and other life. There is much that remains to be discovered. Biologists do not know with any certainty how many species there are or even why some areas, such as the tropics, are richer in biological diversity than others.

It is known that human activities, mostly unintended, threaten biodiversity by altering habitats and introducing non-native species. There is general agreement on the importance of protecting biological diversity, especially since humans depend on the services provided by living organisms and ecosystems. There is less agreement on the best approaches to conservation and how to balance preservation of habitats with meeting human needs.

Energy

The development of modern civilization has been dependent on both the availability and the advancement of energy. We have witnessed a progression from animal and steam power, to the internal combustion engine and electricity generation and to the harnessing of alternative sources of energy. Because of our reliance on energy sources, it is also important to understand the impacts of energy use on the environment. All aspects of energy – the way it is produced, distributed, and consumed – can affect local, regional, and global environments through land use and degradation, air pollution, the acidification of water and soils, and through global climate change via greenhouse gas emissions.

The majority of our current energy stems from fossil fuels such as coal, oil, and natural gas. Coal dominates in our production of electricity, while oil is the world's primary transportation fuel. Natural gas use, which is most commonly as heating fuel, is currently growing faster than other fossil fuels. While cleaner and less carbon intensive than both coal and oil, its use still emits significant amounts of carbon dioxide.

Over the foreseeable future, it is very likely that fossil fuels will remain our largest source of energy. However, fossil fuels are finite resources and there is concern not only about both domestic supply and reliance on foreign supplies, but also with the increasing cost of these fuels. The burning of fossil fuels is also the largest source of carbon dioxide emissions, which contributes to the greenhouse effect. As these fuels continue to decrease in supply and impact the environment, there will be an increasing need to utilize alternative fuel sources.

Nuclear energy provides nearly a fifth of the world's electricity and does not produce harmful by-products, only non-radioactive water vapour. However, controversy remains over the use of nuclear energy due to concerns over safe storage and disposal of radioactive waste, and the potential for accidents and radiation contamination and exposure.

Hydrogen is the most abundant element in the universe and could be an important factor in our energy future. Hydrogen fuel cells can produce power without emitting any pollutants; their only byproducts are water and heat. Hydrogen can both carry and store energy and can be used in a wide variety of applications, including portable devices that use batteries, transportation vehicles, and a number of stationary power sources. The majority of hydrogen is currently produced by processing fossil fuels (which emits pollutants through the initial process), but it can also be manufactured from renewable energy sources and feedstocks through the electrolysis of water.

With the increased cost and environmental impacts related to fossil fuel use, along with the controversy over nuclear power, research and development in the area of renewable sources of energy continues to flourish. Alternative sources – such as wind, solar, geothermal, and water – have been used in one form or another for many centuries, but will require additional advancement before they can become truly cost-competitive with conventional energy sources.

Wind is an abundant, inexhaustible, and clean source of energy; and, advancements in the design of wind turbines has lowered the cost of wind-generated electricity. However, wind variability, noise pollution caused by the turbines, and the potential to harm birds continue to be areas of concern.

Solar technology obtains power from sunlight. There are three types of solar technology: solar electric, solar hot water, and concentrating solar power. The cost of solar power has declined over the past 30 years, and will continue to decline with increased technological efficiency of technology and manufacturing improvements. However, it is also subject to changes in weather (cloud cover) and pollution, which can block sunlight and decrease overall productivity.

Geothermal energy taps into heat underneath the Earth's crust to boil water that is then used to drive electric turbines to heat buildings, homes, or in other non-electrical purposes. Over the past two decades, geothermal energy costs have decreased nearly 50%. However, while abundant, in only a few locations is it close enough to the Earth's surface to provide for large scale power.

Water – or hydroelectric – power has been harnessed for thousands of years, beginning with the water wheel. For more than a century, the technology for using falling water to create hydroelectricity has evolved to its current use of turbines and generators to convert the energy into electricity, which is then fed into the electric grid. Although a widely accessible energy source in many countries – concerns over land use, impacts on fish populations, and water flow and quality remain.

One effective way of reducing environmental impacts from the current production of energy is to increase energy conservation. By using energy more efficiently, both costs and emissions are lowered; therefore, a significant incentive exists to improve efficiency in manufacturing processes, the construction of new buildings, the production of new appliances, and for many other energy uses. Several developed countries are already taking action to increase energy efficiency and, on a larger scale, new measures are being explored to assist developing countries with energy conservation as they continue to increase their energy consumption. Under the Kyoto Protocol, developed countries can provide technological assistance through "clean technology" projects with developing nations and can even put the emission reductions toward compliance in their own countries.

As human consumption of energy increases and we come closer to using up our supply of fossil fuels, further research and development will be necessary to produce alternative and/or renewable sources of energy that are readily available, affordable, and less harmful to the environment. While our dependence on energy is not likely to decrease, there will continue to be an increase of new innovations in energy technologies with a larger focus on energy conservation and efficiency.

Energy Information Administration (EIA): The EIA also provides information on the various types of energy, including renewable, nonrenewable, and secondary sources. Congressional Research Service: Energy Issues Policy papers on energy issues can be found on this site, including articles and several reports related to energy policy and environmental protection.

Food

Human beings began domesticating plants and animals as long as 10,000 years ago. Success in this endeavor led to drastic changes in how and where human beings lived. It changed how humans interacted with each other and how they interacted with the earth. Agriculture led humans from a nomadic existence to one based in permanent and semi-permanent settlements. Human civilization has its roots in the early domestication of plants and animals.

The unprecedented growth in global population that occurred in the twentieth century was made possible by remarkable advances in agriculture and technology. The Green

Revolution brought high-yield crops and advanced growing techniques to developing countries and improved nutrition and health for most parts of the world. Continuing population growth will require more acres of land to plant and more water for irrigation, increasing pressure on habitats and natural resources.

Advanced agricultural methods have made it possible to grow more food on fewer acres of land, which permits much land to be returned to forest and other uses. However, there are costs associated with high-yield methods, including their heavy reliance on pesticides and fertilizers. Runoff of fertilizers from agricultural lands has affected the water quality of some lakes, rivers, and bays. Erosion is, and has been since crops were first planted, a serious environmental problem in the absence of preventative measures. Irrigation, required in many parts of the world, can be a significant drain on water supplies in arid areas and, if improperly managed, can lead to buildup of salt deposits, which may severely degrade soil quality.

One of the most critical challenges of the next few decades is to find ways to increase production of food while minimizing environmental degradation. Considerable research in this and other countries has led to advanced methods to preserve soil and prevent erosion, including conservation tilling and computer-controlled application of pesticides, fertilizers, and water. Pesticide use is controversial. Other methods, especially new types of genetic modification, or "genetic engineering" are more controversial.

To reduce reliance on pesticides, methods such as biocontrol (using other species to control pests) and biotechnology to create pest-resistant crops are being tested. Biocontrol using wild-type (naturally occurring) species is usually accepted, especially by organic producers. Naturally occurring mutants are accepted as well. More controversial is the use of genetically modified biocontrol agents changed by DNA deletion or insertion of new traits. Concerns about the potential impacts of these methods have been raised.

Environment and Society

Throughout history humans have both affected, and been affected by, the natural world. While a good deal has been lost due to human actions, much of what is valued about the environment has been preserved and protected through human action. While many uncertainties remain, there is a realization that environmental problems are becoming more and more complex, especially as issues arise on a more global level, such as that of atmospheric pollution or global warming.

Interactions between human society and the environment are constantly changing. The environment, while highly valued by most, is used and altered by a wide variety of people with many different interests and values. Difficulties remain on how best to ensure the protection of our environment and natural resources. There will always be tradeoffs and, many times, unanticipated or unintended consequences. However, a well-managed environment can provide goods and services that are both essential for our well being as well as for continued economic prosperity.

The environment has become one of the most important issues of our time and will continue to be well into the future. The challenge is to find approaches to environmental management that give people the quality of life they seek while protecting the environmental systems that are also the foundations of our well being. A multidisciplinary approach to learning can build upon the strengths of a wide range of fields of study, providing a deeper understanding of the technological, political, and social options and strategies for both studying and managing the relationship between our society and the environment.

6

Petroleum

PETROLEUM

Petroleum, oily, flammable liquid that occurs naturally in deposits, usually beneath the surface of the earth; it is also called crude oil. It consists principally of a mixture of hydrocarbons

Hydrocarbon

Hydrocarbon any organic compound composed solely of the elements hydrogen and carbon. The hydrocarbons differ both in the total number of carbon and hydrogen atoms in their molecules and in the proportion of hydrogen to carbon. The hydrocarbons can be divided into various homologous series. Each member of such a series shows a definite relationship in its structural formula to the members preceding and following it, and there is generally some regularity in changes in physical properties of successive members of a series. The alkanes are a homologous series of saturated aliphatic hydrocarbons. The first and simplest member of this series is methane, CH_4; the series is sometimes called the methane series. Each successive member of a homologous series of hydrocarbons has one more carbon and two more hydrogen atoms in its molecule than the preceding member. The second alkane is ethane, C_2H_6, and the third is propane, C_3H_8. Alkanes have the general formula C_nH_{2n+2} (where n is an integer greater than or equal to 1). Generally, hydrocarbons of low molecular weight, e.g., methane, ethane, and propane, are gases; those of intermediate molecular weight, e.g., hexane, heptane, and octane, are liquids; and those of high molecular weight, e.g., eicosane ($C_{20}H_{42}$) and polyethylene, are solids. Paraffin is a mixture of high-molecular-weight alkanes; the alkanes are sometimes called the paraffin series. Other homologous series of hydrocarbons include the alkenes alkene (ãl'kçn), any of a group of aliphatic hydrocarbons whose molecules contain one or more carbon-carbon double bonds Many common natural substances, e.g., natural gas, petroleum, and asphalt, are complex mixtures of hydrocarbons. The coal tar obtained from coal by coking is also a mixture of hydrocarbons. Natural gas, petroleum, and coal tar are important sources of many hydrocarbons. These complex mixtures can be refined into simpler mixtures or pure substances by fractional distillation. During the refining of petroleum, one kind of hydrocarbon is often converted to another, more useful kind by cracking. Useful hydrocarbon mixtures include cooking gas, gasoline, naphtha, benzine, kerosene, paraffin, and lubricating oils. Many hydrocarbons are useful as fuels; they burn in air to form carbon dioxide and water.

The hydrocarbons differ in chemical activity. The alkanes are unaffected by many common reagents, while the alkenes and alkynes are much more reactive, as a result of the presence of unsaturation (i.e., a carbon-carbon double or triple bond) in their molecules. Many important compounds are derived from hydrocarbons, either by substitution or replacement by some other chemical group or element of one or more of the hydrogen atoms of the hydrocarbon molecule, or by the addition of some element or group to a double or triple bond (in an unsaturated hydrocarbon). Such derivatives include alcohols, aldehydes, ethers, carboxylic acids, and halocarbons.

Alkane

Alkane any of a group of aliphatic hydrocarbons whose molecules contain only single bonds chemical bond, mechanism whereby atoms combine to form molecules. There is a chemical bond between two atoms or groups of atoms when the forces acting between them are strong enough to lead to the formation of an aggregate with sufficient stability to be regarded as. Alkanes have the general chemical formula C_nH_{2n+2}. An alkane is said to have a continuous chain if each carbon atom in its molecule is joined to at most two other carbon atoms; it is said to have a branched chain if any of its carbon atoms is joined to more than two other carbon atoms.

The first four continuous-chain alkanes are methane, ethane, propane, butane Names of continuous-chain alkanes whose molecules contain more than five carbon atoms are formed from a root that indicates the number of carbon atoms and the suffix *-ane* to indicate that the compound is an alkane; e.g., alkanes with 5, 6, 7, 8, 9, and 10 carbon atoms in their molecules are pentane, hexane, heptane, octane, nonane, and decane, respectively. The name of a branched-chain alkane is formed by adding prefixes to the name of the continuous-chain alkane from which it is considered to be derived; e.g., 2-methylpropane (called also isobutane) is thought of as being derived by replacing one of the hydrogen atoms bonded to the second (2-) carbon atom of a propane molecule with a methyl (CH_3) group, forming $CH_3CH(CH_3)_2$. Chemically, the alkanes are relatively unreactive. They are obtained by fractional distillation from petroleum and are used extensively as fuels. The alkanes are sometimes referred to as the methane series (after the simplest alkane) or as paraffins.

Methane

Methane,CH_4, colourless, odorless, gaseous saturated hydrocarbon; the simplest alkane any of a group of aliphatic hydrocarbons whose molecules contain only single bonds. It is less dense than air, melts at "184°C;, and boils at "161.4°C;. It is combustible and can form explosive mixtures with air. Methane occurs naturally as the principal component of natural gas; it is formed by the decomposition of plant and animal matter. When this decomposition occurs underwater in swamps and marshes, marsh gas is released. The firedamp of coal mines is chiefly methane. In the atmosphere methane is a greenhouse gas, helping to trap infrared radiation, electromagnetic radiation having a wavelength in the range from c.75 × 10"^{6} cm to c.100,000 × 10"^{6} cm (0.000075–0.1 cm).

Methane can be prepared in the labouratory by heating sodium acetate with sodium hydroxide, by the reaction of aluminum carbide with water, by the direct combination of carbon and hydrogen, or by the destructive distillation of coal or wood. As natural gas, methane is widely used for fuel. It is also used for carbonizing steel. It is unaffected by many common chemical reagents but reacts violently with chlorine or fluorine in the presence of light and is therefore important as a starting material for the synthesis of solvents, e.g., methylene chloride, chloroform, and carbon tetrachloride, and of some of the Freon refrigerants.

- Natural gas, natural mixture of gaseous hydrocarbons The composition of natural gas varies in different localities. Its chief component, methane, ethane, propane, butane and other hydrocarbon compounds. Some of the hydrocarbons found in gasoline gasoline or petrol, light, volatile mixture of hydrocarbons for use in the internal-combustion engine and as an organic solvent, obtained primarily by fractional distillation and "cracking" of petroleum, but also obtained from natural gas, by also occur as vapours in natural gas; by liquefying these hydrocarbons, gasoline can be obtained. Although commonly associated with petroleum petroleum, oily, flammable liquid that occurs naturally in deposits, usually beneath the surface of the earth; it is also called crude oil. It consists principally of a mixture of hydrocarbons, with traces of various nitrogenous and sulfurous compounds.. deposits it also occurs separately in sand, sandstone, and limestone deposits. Some geologists theorize that natural gas is a byproduct of decaying vegetable matter in underground strata, while others think it may be primordial gases that rise up from the mantle. Because of its flammability and high calorific value, natural gas is used extensively as an illuminant and a fuel.Natural gas was known to the ancients but was considered by them to be a supernatural phenomenon because, noticed only when ignited, it appeared as a mysterious fire bursting from the ground. One of the earliest attempts to harness it for economic use occurred in the early 19th cent. in Fredonia, N.Y. Toward the latter part of the 19th cent., large industrial cities began to make use of natural gas, and extensive pipeline systems have been constructed to transport gas.

- Liquefied natural gas, or LNG, is natural gas that has been pressurized and cooled so as to liquefy it for convenience in shipping and storage. The boiling point of natural gas is extremely low, and only in the 1970s did cryogenic technology physics low-temperature physics, science concerned with the production and maintenance of temperatures much below normal, down to almost absolute zero, and with various phenomena that occur only at such temperatures.advance enough to make the production and transport of LNG commerically feasible. Some of the natural gas moved to and from the United States is carried as LNG in special tankers.
- Infrared radiation, electromagnetic radiation, energy radiated in the form of a wave as a result of the motion of electric charges. A moving charge gives rise to a magnetic field, and if the motion is changing (accelerated), then the magnetic field varies and in turn produces an having a wavelength in the range from c.75 $\times 10^{-6}$ cm to c.100,000 $\times 10^{-6}$ cm (0.000075–0.1 cm). Infrared rays thus occupy that part of the electromagnetic spectrum arrangement or display of light or other form of radiation separated according to wavelength, frequency, energy, or some other property. Beams of charged particles can be separated into a spectrum according to mass in a mass spectrometer with a frequency less than that of visible light visible electromagnetic radiation. Of the entire electromagnetic spectrum, the human eye is sensitive to only a tiny part, the part that is called light. The wavelengths of visible light range from about 350 or 400 nm to about 750 or 800 nm.and greater than that of most radio waves, although there is some overlap. The name infrared means "below the red," i.e., beyond the red, or lower-frequency (longer wavelength), end of the visible

spectrum. Infrared radiation is thermal, or heat, no mechanical energy in transit, associated with differences in temperature between a system and its surroundings or between parts of the same system.

- Measures of Heat: It was first discovered in 1800 by Sir William Herschel, who was attempting to determine the part of the visible spectrum with the minimum associated heat in connection with astronomical observations he was making. In 1847, A. H. L. Fizeau and J. B. L. Foucault showed that infrared radiation has the same properties as visible light, being reflected, refracted, and capable of forming an interference interferometer. When the wavelength of the light is known, the interferometer indicates the thickness of the film by the interference patterns it forms. The reverse process, i.e., the measurement of the length of an unknown light wave, can also be carried out by the interferometer.
- Pattern: Infrared radiation is typically produced by objects whose temperature measure of the relative warmth or coolness of an object. Temperature is measured by means of a thermometer or other instrument having a scale calibrated in units called degrees. The size of a degree depends on the particular temperature scale being used.

There are many applications of infrared radiation. A number of these are analogous to similar uses of visible light. Thus, the spectrum of a substance in the infrared range can be used in chemical analysis much as the visible spectrum is used. Radiation at discrete wavelengths in the infrared range is characteristic of many molecules. The temperature of a distant object can also be determined by analysis of the infrared radiation from the object. Radiometers operating in the infrared range serve as the basis for many instruments, including heat-seeking devices in missiles and devices for spotting and photographing persons and objects in the dark or in fog. Medical uses of infrared radiation range from the simple heat lamp to the technique of thermal imaging, or thermography. A thermograph of a person can show areas of the body where the temperature is much higher or lower than normal, thus indicating some medical problem. Thermography has also been used in industry and other applications. Some lasers produce infrared radiation. A recent development has been the expansion of research in infrared astronomy study of celestial objects by means of the infrared radiation they emit, in the wavelength range from about 1 micrometer to about 1 millimeter. All objects, from trees and buildings on the earth to distant galaxies, emit infrared (IR) radiation.

Infrared sensors are sent aloft in balloons, rockets, and satellites to study the infrared radiation reaching the earth from other parts of the solar system and beyond.electromagnetic radiation, energy in physics, the ability or capacity to do work or to produce change. Forms of energy include heat, light, sound, electricity, and chemical energy.. radiated in the form of a wave, in physics, the transfer of energy by the regular vibration, or oscillatory motion, either of some material medium or by the variation in magnitude of the field vectors of an electromagnetic field as a result of the motion of electric charges. A moving charge gives rise to a magnetic field, and if the motion is changing (accelerated), then the magnetic field varies and in turn produces an electric field.

These interacting electric and magnetic fields are at right angles to one another and also to the direction of propagation of the energy. Thus, an elcctromagnetic wave is a transverse wave. If the direction of the electric field is constant, the wave is said to be polarized orientation of the vibration pattern of light waves in a singular plane.

CHARACTERISTICS OF POLARIZATION

Polarization is a phenomenon peculiar to transverse waves, i.e. Electromagnetic radiation does not require a material medium and can travel through a vacuum. The theory of electromagnetic radiation was developed by James Clerk Maxwell and published in 1865. He showed that the speed of propagation of electromagnetic radiation should be identical with that of light visible electromagnetic radiation.

Of the entire electromagnetic spectrum, the human eye is sensitive to only a tiny part, the part that is called light. The wavelengths of visible light range from about 350 or 400 nm to about 750 or 800 nm.. Subsequent experiments by Heinrich Hertz verified Maxwell's prediction through the discovery of radio waves, also known as hertzian waves.

Light is a type of electromagnetic radiation, occupying only a small portion of the possible spectrum arrangement or display of light or other form of radiation separated according to wavelength, frequency, energy, or some other property. Beams of charged particles can be separated into a spectrum according to mass in a mass spectrometer of this energy. The various types of electromagnetic radiation differ only in wavelength and frequency; they are alike in all other respects.

The possible sources of electromagnetic radiation are directly related to wavelength: long radio waves are produced by large antennas such as those used by broadcasting stations; much shorter visible light waves are produced by the motions of charges within atoms, basic unit of matter more properly, the smallest unit of a chemical element having the properties of that element.

STRUCTURE OF THE ATOM

The shortest waves, those of gamma radiation, high-energy photons emitted as one of the three types of radiation resulting from natural radioactivity. It is the most energetic form of electromagnetic radiation, with a very short wavelength (high frequency).result from changes within the nucleus, in physics, the extremely dense central core of an atom.

THE NATURE OF THE NUCLEUS

Composition

Atomic nuclei are composed two types of particles, protons and neutrons, which are collectively known as nucleons. of the atom. In order of decreasing wavelength and increasing frequency, various types of electromagnetic radiation include: electric waves, radio, transmission or reception of electromagnetic radiation in the radio frequency range. The term is commonly applied also to the equipment used, especially to the radio receiver.waves (including AM, FM, TV, and shortwaves), microwaves, infrared radiation infrared radiation, electromagnetic radiation having a wavelength in the range from c.75 × 10^{-6} cm to c.100,000 × 10^{-6} cm (0.000075–0.1 cm).

visible light, ultraviolet radiation UV index predicts how long it would take a light-skinned American to get a sunburn if exposed, unprotected, to the noonday sun, given the geographical location and the local weather., X rays, invisible, highly penetrating electromagnetic radiation of much shorter wavelength (higher frequency) than visible light. The wavelength range for X rays is from about 10^{-8} m to about 10^{-11}, and gamma radiation. According to the quantum theory quantum theory, modern physical theory concerned with the emission and absorption of energy by matter and with the motion of material particles; the quantum theory and the theory of relativity together form the theoretical basis of modern physics. light and other forms of electromagnetic radiation may at times exhibit properties like those of particles in their interaction with matter. The individual quantum of electromagnetic radiation is known as the photon, the particle composing light and other forms of electromagnetic radiation, sometimes called light quantum and is symbolized by the Greek letter gamma. Quantum effects are most pronounced for the higher frequencies, such as gamma rays, and are usually negligible for radio waves at the long-wavelength, low-frequency end of the spectrum.

Propane,

Propane, $CH_3CH_2CH_3$, colourless, gaseous alkane, any of a group of aliphatic hydrocarbons whose molecules contain only single bonds It is readily liquefied by compression and cooling. It melts at “189.9°C; and boils at “42.2°C;. Propane occurs in nature in natural gas and (in dissolved form) in crude oil; it is also a byproduct of petroleum refining. It is chiefly as a fuel. For this purpose it is sold compressed in cylinders of various sizes, often mixed with other hydrocarbons, e.g., butane. Propane fuel is used in a type of cigarette lighter and in portable stoves and lamps

Butane

Butane, C_4H_{10}, gaseous alkane), any of a group of aliphatic hydrocarbons whose molecules contain only single bonds, a hydrocarbon that is obtained from natural gas or by refining petroleum. It can be liquefied at room temperature by compression. There are two structural isomers, in chemistry, one of two or more compounds having the same molecular formula but different structures. of butane. In normal butane, or *n*-butane, the four carbon atoms are joined in a continuous, unbranched chain; in isobutane, or 2-methylpropane, three of the carbon atoms are joined to the fourth by single bonds, resulting in a branched structure. The two isomers differ in certain of their chemical and physical properties, e.g., liquid *n*-butane has a higher boiling point ("0.6°C;) at atmospheric pressure than that of liquid isobutane ("10.2°C;).

Benzene

Benzene colourless, flammable, toxic liquid with a pleasant aromatic odor. It boils at 80.1& deg C; and solidifies at 5.5 & deg C;. Benzene is a hydrocarbon hydrocarbon, any organic compound composed solely of the elements hydrogen and carbon.., with formula C_6H_6. The simplest picture of the benzene molecule, proposed by the German chemist, is a hexagon of six carbon atoms joined by alternating single and double bonds and each bearing one hydrogen atom, symbolized by [N/A]. However, modern studies have shown that the six carbon-carbon bonds are all of equal strength and distance; thus the double-bond electrons do not belong to any particular bonds but rather are delocalized about the ring, with the result that the strength of each bond is between that of a single bond and that of a double bond.

There is a chemical bond between two atoms or groups of atoms when the forces acting between them are strong enough to lead to the formation of an aggregate with sufficient stability to be regarded.Benzene is the parent substance of the aromatic compounds aromatic compound, any of a large class of compounds that includes benzene and compounds that resemble benzene in certain of their chemical properties. Originally applied to a small class of pleasant-smelling chemicals derived from vegetables, it now encompasses a. large and important group of organic compounds. It is the first of a series of hydrocarbons known as the benzene series, formed by the substitution of methyl groups, CH_3, for the hydrogen atoms of the benzene molecule. The second member of the series is toluene,$C_6H_5CH_3$, from which trinitrotoluene trinitrotoluene. is derived, and the third member is xylene xylene or dimethylbenzene., $C_6H_4(CH_3)_2$, a solvent.

In xylene and other benzene derivatives in which two of the hydrogens have been replaced, there are three possible arrangements of the substitution groups; in the *ortho* (*o*) configuration the groups are on adjacent carbon atoms, in the *meta* (*m*) configuration the groups are separated by one carbon atom, and in the *para* (*p*) configuration the groups are on opposite sides of the ring. In addition to derivatives formed by the substitution of other groups for one or more of the hydrogen atoms of the benzene ring, two or more rings may be joined together, as in naphthalene, colourless, crystalline, solid aromatic hydrocarbon with a pungent odor., anthracene, $C_{14}H_{10}$, solid organic compound derived from coal tar.and phenanthrene; or other atoms, such as nitrogen, may be substituted for carbon atoms in the ring, as in pyridine (C_5H_5N) and pyrimidine ($C_4H_4N_2$). Among the important derivatives of benzene are phenol,C_6H_5., aniline, $C_6H_5NH_2$., and picric acid picric acid or 2,4,6-trinitrophenol.

Benzene and the other aromatic hydrocarbons are obtained for industrial purposes from the distillation of coal tar, a byproduct in the manufacture of coke, and from petroleum by special reforming methods. They are used in the manufacture of plastics, synthetic rubber, dyes, and drugs. Benzene is a known carcinogen.

Toluene

Toluene or methylbenzene C_7H_8, colourless liquid aromatic hydrocarbon that melts at "95°C; and boils at 110.8°C;. It is insoluble in water but highly soluble in most organic solvents. Toluene is obtained from coal tar and petroleum by distillation. It is used as a solvent and as a starting material for the synthesis of many compounds, including dyes and explosives. When toluene is treated with a mixture of nitric and sulfuric acids (a process known as nitration), trinitrotoluene

Xylene

Xylene used extensively as a solvent, obtained from coal tar, wood tar, and sometimes from petroleum. It is a mixture of three isomers that differ structurally from one another in the location of the two methyl groups that have replaced hydrogen atoms in the benzene. molecule. *Ortho*-xylene is 1,2-dimethylbenzene; it melts at "25°C; and boils at 144°C;. *Meta*-xylene is 1,3-dimethylbenzene; it melts at "48°C; and boils at 139°C;. *Para*-xylene is 1,4-dimethylbenzene; it melts at 13°C; and boils at 138°C;. The separation of these three isomers from one another by fractional distillation is difficult because their boiling points are so close together. The *ortho* and *para* isomers are converted to *meta*-xylene by treatment with aluminum trichloride and hydrochloric acid at about 80°C;. The xylenes are often used in the synthesis of other compounds, e.g., the xylidenes that are amino derivatives used in the synthesis of azo dyes and other compounds.

Pyridine

Pyridine C_5H_5N, colourless, flammable, toxic liquid with a putrid odor. It melts at "42°C; and boils at 115.5°C;. Chemically, it is a heterocyclic aromatic tertiary amine under amino group amine. Amines are derivatives of the inorganic compound ammonia, NH_3. When one, two, or all three of the hydrogens in ammonia are replaced by an alkyl or aryl group, the resulting compound is known as a primary, secondary, or tertiary amine. Its molecule resembles that of benzene, one carbon-hydrogen unit in the benzene ring being replaced with a nitrogen molecule. It is miscible with water and with most organic solvents. Its aqueous solution is slightly alkaline. Pyridine is used as a solvent, as a denaturant for alcohol, and as a starting material in the synthesis of other compounds. Compounds that can be derived from pyridine include antihistamines and vitamins. Pyridine is obtained from bone oil or from coal tar by destructive distillation, which decomposes alkaloids that contain it. Alkaloids that contain pyridine include coniine, piperine (the alkaloid in pepper), and nicotine (present in tobacco); free pyridine is present in tobacco smoke.

Pcric Acid

Picric acid or 2,4,6-trinitrophenol, $C_6H_2(NO_2)_3OH$, a toxic yellow crystalline solid that melts at 122°C; and is soluble in most organic solvents. Picric acid is a derivative of phenol, C_6H_5. It reacts with metals to form metal picrates, which like picric acid itself are highly sensitive explosives substance that undergoes decomposition or combustion with great rapidity, evolving much heat and producing a large volume of gas. The reaction products fill a much greater volume than that occupied by the original material and exert an enormous pressure,. that can be detonated by heat, flame, shock, or friction. The high explosives lyddite and melinite are composed mostly of compressed or fused picric acid. Picric acid is often used as a booster to detonate another, less sensitive explosive, such as TNT (trinitrotoluene. Although picric acid can be synthesized by nitration of phenol, higher yields are obtained if chlorobenzene is used as a starting material; the latter method involves several steps and the formation of several intermediate products. In addition to its use in explosives, picric acid has been used as a yellow dye, as an antiseptic, and in the synthesis of chloropicrin, or nitrotrichloromethane, CCl_3NO_2, a powerful insecticide.

ORIGIN AND NATURAL OCCURRENCE

During the past 600 million years incompletely decayed plant and animal remains have become buried under thick layers of rock. It is believed that petroleum consists of the remains of these organisms but it is the small microscopic plankton organism remains that are largely responsible for the relatively high organic carbon content of fine-grained sediments like the Chattanooga shale which are the principle source rocks for petroleum. Among the leading producers of petroleum are Saudi Arabia, Russia, the United States (chiefly Texas, California, Louisiana, Alaska, Oklahoma, and Kansas), Iran, China, Norway, Mexico, Venezuela, Iraq, Great Britain, the United Arab Emirates, Nigeria, and Kuwait. The largest known reserves are in the Middle East.

EXPLORATION AND DRILLING OF WELLS

Because of the subterranean origin of petroleum it must be extracted by means of wells well, aperture in the earth's surface through which substances in a natural underground reservoir, such as water, gas, oil, salt, and sulfur, can flow or be pumped to the surface.

Until an exploratory well, or wildcat, has been dug, there is no sure way of knowing whether or not petroleum lies under a particular site. In order to reduce the number of exploratory wells drilled, scientific methods are used to pick the most promising sites. Sensitive instruments, such as the gravimeter, the magnetometer, and the seismograph, may be used to find subsurface rock formations that can hold crude oil. Drilling is a fairly complex and often risky process. Some wells must be dug several miles deep before petroleum deposits are reached. Many are now drilled offshore from platforms standing in the ocean bed. Usually the petroleum from a new well will come to the surface under its own pressure. Later the crude oil must be pumped out or forced to the surface by injecting water, air, natural gas, steam, carbon dioxide, or another substance into the deposits. Enhanced recovery techniques have increased the percentage of oil that can be extracted from a field.

Well

Well, aperture in the earth's surface through which substances in a natural underground reservoir, such as water, gas, oil, salt, and sulfur, can flow or be pumped to the surface. In the United States, until some years after the Civil War, the majority of wells were "open," i.e., holes dug in the ground and lined, or cased, with brick, stone, or wood. Although they are sometimes dug with picks and shovels, most wells today are made by rotary or percussion drills. An artesian well artesian well, deep drilled well through which water is forced upward under pressure. The water in an artesian well flows from an aquifer, which is a layer of very porous rock or sediment, usually sandstone, capable of holding and transmitting large quantities of the most desirable type of water well, is always drilled because rock layers must be cut through to reach the water. Oil wells are usually drilled using a rotary-drill method, in which a drilling bit set in the bottom of a drilling pipe is rotated by machinery on the ground level. As the cut deepens, more sections of pipe are fastened to the sections already in use. A special mixture called drilling mud is sent down through the pipe to wash away the drillings and also to cool the cutting bit. Some oil wells are drilled by a percussion method known as cable-tool drilling. In this procedure a heavy metal bit attached to a cable is alternately raised and dropped, pulverizing the rock beneath it. Water is pumped into the well and mixed with the rock cuttings, the mixture being bailed out when it becomes thick enough to interfere with the action of the bit. Regardless of the drilling method, well walls are usually cased with iron or steel to prevent cave-ins. Casing is inserted when the desired depth has been reached or, in some instances, as the well is being drilled. Minerals, such as salt and sulfur, can be pumped to the surface through a well if they are first liquefied by some process; salt may be brought up if water is first pumped to the bottom of the well to dissolve the salt.

Artesian Well

Artesian well, deep drilled well well, aperture in the earth's surface through which substances in a natural underground reservoir, such as water, gas, oil, salt, and sulfur, can flow or be pumped to the surface through which water is forced upward under pressure. The water in an artesian well flows from an aquifer, which is a layer of very porous rock or sediment, usually sandstone, capable of holding and transmitting large quantities of water. The geologic conditions necessary for an artesian well are an inclined aquifer sandwiched between impervious rock layers above and below that trap water in it. Water enters the exposed edge of the aquifer at a high elevation and percolates downward through interconnected pore spaces. The water held in these spaces is under pressure because of the weight of water in the portion of the aquifer above it. If a well is drilled from the land surface through the overlying impervious layer into the aquifer, this pressure will cause the water to rise in the well. In areas where the slope of the aquifer is great enough, pressure will drive the water above ground level in a spectacular, permanent fountain. Artesian springs can occur in similar fashion where faults or cracks in the overlying impervious layer allow water to flow upward. Water from an artesian well or spring is usually cold and free of organic contaminants, making it desirable for drinking. In North America, the Dakota sandstone provides aquifers for an artesian system that underlies parts of the Dakotas, Montana, Wyoming, Kansas, Nebraska, and Saskatchewan and supplies great quantities of water to the dry Great Plains region. Many East Coast cities derive their water supplies from aquifers that are exposed along the edge of the Piedmont and dip downward toward the Atlantic coast. The largest artesian system in the world underlies nearly all of E and S Australia. Other important artesian systems serve London, Paris, and E Algeria.

COMPOSITION AND REFINING OF PETROLEUM

The physical properties and exact chemical composition of crude oil varies from one locality to another. The different hydrocarbon components of petroleum are dissolved natural gas Liquefied natural gas, or LNG, is natural gas that has been pressurized and cooled so as to liquefy it for convenience in shipping and storage. The boiling point of natural gas is extremely low, and only in the 1970s did cryogenic technology advance, gasoline, benzine, colourless, highly flammable liquid.

Naphtha, kerosene or kerosine, colourless, thin mineral oil whose density is between 0.75 and 0.85 grams per cubic centimeter. A mixture of hydrocarbons, it is commonly obtained in the fractional distillation of petroleum as the portion boiling off., diesel fuel and light heating oils, heavy heating oils, and finally tars of various weights tar and pitch, viscous, dark-brown to black substances obtained by the destructive distillation of coal, wood, petroleum, peat, and certain other organic materials.. The crude oil is usually sent from a well to a refinery in pipelines, hollow structure, usually cylindrical, for conducting materials. It is used primarily to convey liquids, gases, or solids suspended in a liquid, e.g., a slurry. It is also used as a conduit for electric wires.

Or tanker ships.The hydrocarbon components are separated from each other by various refining processes. In a process called fractional distillation distillation, process used to separate the substances composing a mixture. It involves a change of state, as of liquid to gas, and subsequent condensation. The process was probably first used in the production of intoxicating beverages.. petroleum is heated and sent into a tower. The vapours of the different components condense on collectors at different heights in the tower. The separated fractions are then drawn from the collectors and further processed into various petroleum products.

One of the many products of crude oil is a light substance with little colour that is rich in gasoline or petrol, light, volatile mixture of hydrocarbons for use in the internal-combustion engine and as an organic solvent, obtained primarily by fractional distillation and "cracking" of petroleum, but also obtained from natural gas, by. Another is a black tarry substance that is rich in asphalt brownish-black substance used commonly in road making, roofing, and waterproofing.As the lighter fractions, especially gasoline, are in the greatest demand, so-called cracking processes have been developed in which heat, pressure, and certain catalysts are used to break up the large molecules of heavy hydrocarbons into small molecules of light hydrocarbons. Some of the heavier fractions find eventual use as lubricating oils, paraffins white, more-or-less translucent, odorless, tasteless, waxy solid. It melts between 47°C; and 65°C; and is insoluble in water but soluble in ether, benzene, and certain esters., and highly refined medicinal substances such as petrolatum..

Benzine

Benzine colourless, highly flammable liquid. It is used as a cleaning agent because it is a solvent for organic substances such as fats, oils, and resins and is also used in the preparation of certain dyes and paints. Benzine is a mixture of hydrocarbons, chiefly alkanes such as pentane and hexane. It is obtained by the fractional distillation of petroleum oily, flammable liquid that occurs naturally in deposits, usually beneath the surface of the earth; it is also called crude oil. It consists principally of a mixture of hydrocarbons, with traces of various nitrogenous and sulfurous compounds.

Naphtha

Naphtha, term usually restricted to a class of colourless, volatile, flammable liquid hydrocarbon mixtures. Obtained as one of the more volatile fractions in the fractional distillation of petroleum petroleum, oily, flammable liquid that occurs naturally in deposits, usually beneath the surface of the earth; it is also called crude oil. It consists principally of a mixture of hydrocarbons, with traces of various nitrogenous and sulfurous compounds. (when it is known as petroleum naphtha), in the fractional distillation of coal tar (coal-tar naphtha), and in a similar distillation of wood (wood naphtha), it is used widely as a solvent for various organic substances, such as fats and rubber, and in the making of varnish. Because of its dissolving property it is important as a cleaning fluid; it is also incorporated in certain laundry soaps. Coal-tar (aromatic) naphthas have greater solvent power than petroleum (aliphatic) naphthas. Originally the term *naphtha* designated a colourless flammable liquid obtained from the ground in Persia. Later it came to be applied to a number of other natural liquid substances having similar properties. Technically, gasoline and kerosene are considered naphthas.

Kerosene or Kerosine

Kerosene or kerosine, colourless, thin mineral oil whose density is between 0.75 and 0.85 grams per cubic centimeter. A mixture of hydrocarbons, it is commonly obtained in the fractional distillation of petroleum as the portion boiling off between 150°C; and 275°C; (302°F;–527°F;). Kerosene has been recovered from other substances, notably coal (hence another name, coal oil), oil shale, and wood. At one time kerosene was the most important refinery product because of its use in lamps. Now it is most noted for its use as a carrier in insecticide sprays and as a fuel in jet engines.

Paraffin

Paraffin, white, more-or-less translucent, odorless, tasteless, waxy solid. It melts between 47°C; and 65°C; and is insoluble in water but soluble in ether, benzene, and certain esters. Paraffin is unaffected by most common chemical reagents but burns readily in air. Obtained from petroleum during refining, it is used in candles, for coating paper, and for various other purposes. Chemically, paraffin is a mixture of high-molecular-weight alkanes, any of a group of aliphatic hydrocarbons whose molecules contain only single bonds i.e., saturated hydrocarbons any organic compound composed solely of the elements hydrogen and carbon.

With the general formula C_nH_{2n+2}, where *n* is an integer between 22 and 27.

Petrolatum

Petrolatum colourless to yellowish-white hydrocarbon mixture obtained by fractional distillation of petroleum. In its jellylike semisolid form (known as petroleum jelly and also by several trade names) it is used in preparing medicinal ointments and for lubrication. As a nearly colourless, highly refined liquid known as liquid petrolatum, liquid paraffin, or mineral oil, it is used as a lubricant, as a laxative, and as a base for nasal sprays.

Asphalt

Asphalt brownish-black substance used commonly in road making, roofing, and waterproofing. Chemically, it is a natural mixture of hydrocarbons. It varies in consistency from a solid to a semisolid, has great tenacity, melts when heated, and when ignited will burn with a smoky flame leaving very little or no ash. It is found in nature in deposits called asphalt lakes. Natural asphalt was probably formed by the evapouration of petroleum. Asphalt is obtained as a residue in the distillation or refining of petroleum. This is its important commercial source. It occurs also in asphalt rock, a natural mixture of asphalt with sand and limestone, which when crushed is used as road-building material. Asphalt is also used in the manufacture of paints and varnishes, giving an intensely black colour.

Gasoline or Petrol

Gasoline or petrol, light, volatile mixture of hydrocarbons for use in the internal-combustion engine internal-combustion engine, one in which combustion of the fuel takes place in a confined space, producing expanding gases that are used directly to provide mechanical power. and as an organic solvent, obtained primarily by fractional distillation and "cracking" of petroleum, but also obtained from natural gas, by destructive distillation of oil shales and coal, and by a process that converts methanol to gasoline using zeolite as a catalyst. Gasoline intended for use in engines is rated by octane number octane number, figure of merit representing the resistance of gasoline to premature detonation when exposed to heat and pressure in the combustion chamber of an internal-combustion engine., an index of quality that reflects the ability of the fuel to resist detonation and burn evenly when subjected to high pressures and temperatures inside an engine. Premature detonation produces "knocking" and "pinging"; it wastes fuel and may cause engine damage.

The addition of tetramethyl lead and tetraethyl lead to raise the octane number is no longer permitted in the United States because it leads to dangerous emissions containing lead. New formulations of gasoline designed to raise the octane number contain increasing amounts of aromatics and oxygen-containing compounds (oxygenates), such as alcohols, methyl tertiary butyl ether (MTBE), and methylcyclopentadienyl manganese tricarbonyl (MMT).

Automobiles automobile, self-propelled vehicle used for travel on land. The term is commonly applied to a four-wheeled vehicle designed to carry two to six passengers and a limited amount of cargo, as contrasted with a truck, which is designed primarily for the transportation are now equipped with catalytic converters that oxidize unreacted gasoline; the cars are designed to run on newly formulated gasolines as well as on gasohol gasohol, a gasoline extender made from a mixture of gasoline (90%) and ethanol (10%; often obtained by fermenting agricultural crops or crop wastes) or gasoline (97%) and methanol, or wood alcohol (3%)., which contains 10% ethanol or 3% methanol. In addition, since 1998 a number of American automobiles have been equipped to enable them to run on either gasoline or E85, a mixture of 85% ethanol and 15% gasoline. Some racing cars use pure methanol methanol, methyl alcohol, or wood alcohol, CH_3OH, a colourless, flammable liquid that is miscible with water in all proportions. Methanol is a monohydric alcohol

There are five blends of gasoline marketed in the United States. *Conventional gasoline,* the most widely available, is sold where air quality is satisfactory; it has been formulated to evapourate more slowly in hot weather so as to reduce smog, and it now contains detergent additives to reduce engine deposits. *Winter oxygenated gasoline,* is formulated as conventional gasoline with oxygen-rich chemicals added, such as MTBE or ethanol.

The oxygen promotes cleaner burning, reducing carbon monoxide, and is generally sold from November to March because cold engines operate less efficiently and produce more carbon monoxide. *Reformulated gasoline* (RFG), is mandated in areas where toxins in the air are a constant problem; it contains oxygen-rich chemicals in lesser concentrations than the winter oxygenated gasoline and is formulated to reduce certain toxic chemicals found in conventional and winter oxygenated fuels. *Oxygenated reformulated gasoline* is a wintertime fuel exclusive to the New York City area, where heavy carbon monoxide pollution occurs. *California reformulated gasoline,* introduced has a different formulation and burns cleaner than regular reformulated gasoline. Because MTBE has been implicated as a pollutant, particularly of groundwater, its use is being curtailed. California ruled that the MTBE in California reformulated gas must be phased.

Distillation

Distillation, process used to separate the substances composing a mixture. It involves a change of state, as of liquid to gas, and subsequent condensation condensation, in physics, change of a substance from the gaseous (vapour) to the liquid state Condensation is the reverse of vapourization, or change from liquid to gas.. The process was probably first used in the production of intoxicating beverages. Today, refined methods of distillation are used in many industries, including the alcohol and petroleum industries.

Condensation

Condensation, in physics, change of a substance from the gaseous (vapour) to the liquid state, forms of matter differing in several properties because of differences in the motions and forces of the molecules of which they are composed.

Vapourization

Vapourization, change of a liquid or solid substance to a gas or vapour. There is fundamentally no difference between the terms *gas* and *vapour,* but gas is used commonly to describe a substance that appears in the gaseous state under standard conditions of or change from liquid to gas. It can be brought about by cooling, as in distillation, process used to separate the substances composing a mixture. It involves a change of state, as of liquid to gas, and subsequent condensation. The process was probably first used in the production of intoxicating beverages.or by an increase in pressure resulting in a decrease in volume. Certain natural phenomena, such as dew, fog, mist, and clouds, are the result of the condensation of water vapour in the atmosphere; the formation of dew, thin film of water that has condensed on the surface of objects near the ground. Dew forms when radiational cooling of these objects during the nighttime hours also cools the shallow layer of overlying air in contact with them, causing the condensation of some. illustrates well the fundamental principles involved in such phenomena. The explanation of condensation can be found in the kinetic-molecular theory of gases, physical theory that explains the behaviour of gases on the basis of the following assumptions:

- Any gas is composed of a very large number of very tiny particles called molecules;
- The molecules are very far apart compared to. As heat is removed from a gas, the molecules of the gas move more slowly

The intermolecular forces that are exerted by molecules on each other and that, in general, affect the macroscopic properties of the material of which the molecules are a part. Such forces may be either attractive or repulsive in nature.

Are strong enough to pull the molecules together to form droplets of liquid. Similarly, reducing the volume of the gas reduces the average distance between molecules and thus favours the intermolecular forces tending to pull them together.

Vapourization, change of a liquid or solid substance to a gas or vapour. There is fundamentally no difference between the terms *gas* and *vapour*, but gas is used commonly to describe a substance that appears in the gaseous state under standard conditions of pressure and temperature, and vapour to describe the gaseous state of a substance that appears ordinarily as a liquid or solid. Although most substances undergo changes of state in the order of solid to liquid to gas as the temperature is raised, a few change directly from solid to gas in a process known as sublimation sublimation (smbl-mâ'shYn), change of a solid substance directly to a vapour without first passing through the liquid state.

THE BOILING POINT AND LATENT HEAT OF VAPOURIZATION

When heat is added to a liquid at its boiling point boiling point, temperature at which a substance changes its state from liquid to gas. A stricter definition of boiling point is the temperature at which the liquid and vapour (gas) phases of a substance can exist in equilibrium., with the pressure kept constant, the molecules of the liquid acquire enough energy to overcome the intermolecular forces intermolecular forces, forces that are exerted by molecules on each other and that, in general, affect the macroscopic properties of the material of which the molecules are a part. Such forces may be either attractive or repulsive in nature.

that bind them together in the liquid state, and they escape as individual molecules of vapour until the vapourization is complete. Vapourization at the boiling point is known simply as boiling. The temperature of a boiling liquid remains constant until all of the liquid has been converted to a gas.

For each substance a certain specific amount of heat must be supplied to vapourize a given quantity of the substance. This amount of heat is known as the latent heat latent heat, heat change associated with a change of state or phase Latent heat, also called heat of transformation, is the heat given up or absorbed by a unit mass of a substance as it changes from a solid to a liquid, from a liquid to a of vapourization of the substance. The quantity of heat applied for each gram (or each molecule) undergoing the change in state depends on the substance itself. The amount of heat necessary to change one gram of water to steam at its boiling point at one atmosphere of pressure, i.e., the heat of vapourization of water, is approximately 540 calories. Other substances require other amounts.

Boiling point, temperature at which a substance changes its state from liquid to gas. A stricter definition of boiling point is the temperature at which the liquid and vapour (gas) phases of a substance can exist in equilibrium. When heat is applied to a liquid, the temperature of the liquid rises until the vapour pressure vapour pressure, pressure exerted by a vapour that is in equilibrium with its liquid. A liquid standing in a sealed beaker is actually a dynamic system: some molecules of the liquid are evapourating to form vapour and some molecules of vapour are condensing to form liquid.. of the liquid equals the pressure of the surrounding gases. At this point there is no further rise in temperature, and the additional heat energy supplied is absorbed as latent heat latent heat, heat change associated with a change of state or phase Latent heat, also called heat of transformation, is the heat given up or absorbed by a unit mass of a substance as it changes from a solid to a liquid, from a liquid to a. vapourization to transform the liquid into gas.

This transformation occurs not only at the surface of the liquid (as in the case of evapouration change of a liquid into vapour at any temperature below its boiling point. For example, water, when placed in a shallow open container exposed to air, gradually disappears, evapourating at a rate that depends on the amount of surface exposed, the humidity but also throughout the volume of the liquid, where bubbles of gas are formed. The boiling point of a liquid is lowered if the pressure of the surrounding gases is decreased. Water will boil at a lower temperature at the top of a mountain, where the atmospheric pressure on the water is less, than it will at sea level, where the pressure is greater.

In the labouratory, liquids can be made to boil at temperatures far below their normal boiling points by heating them in vacuum flasks under greatly reduced pressure. On the other hand, if the pressure is increased, the boiling point is raised. For this reason, it is customary when the boiling point of a substance is given to include the pressure at which it is observed, if that pressure is other than standard, i.e., 760 mm of mercury or 1 atmosphere. The boiling point of a solution solution, in chemistry, homogeneous mixture of two or more substances. The dissolving medium is called the solvent, and the dissolved material is called the solute. A solution is distinct from a colloid or a suspension is always higher than that of the pure solvent; this boiling-point elevation is one of the colligative properties colligative properties, properties of a solution that depend on the number of solute particles present but not on the chemical properties of the solute. Colligative properties of a solution include freezing point.

EVAPOURATION AND VAPOUR PRESSURE

Liquids can also change to gases at temperatures below their boiling points. Vapourization of a liquid below its boiling point is called evapouration evapouration, change of a liquid into vapour at any temperature below its boiling point. Water, when placed in a shallow open container exposed to air, gradually disappears, evapourating at a rate that depends on the amount of surface exposed, the humidity.

Which occurs at any temperature when the surface of a liquid is exposed in an unconfined space. When, however, the surface is exposed in a confined space and the liquid is in excess of that needed to saturate the space with vapour, an equilibrium is quickly reached between the number of molecules of the substance going off from the surface and those returning to it. A change in temperature upsets this equilibrium; a rise in temperature, increases the activity of the molecules at the surface and consequently increases the rate at which they fly off. When the temperature is maintained at the new point for a short time, a new equilibrium is soon established.

The pressure exerted by the vapour of a liquid in a confined space is called its vapour pressure. It differs for different substances at any given temperature, but each substance has a specific vapour pressure for each given temperature. At its boiling point the vapour pressure of a liquid is equal to atmospheric pressure. The vapour pressure of water, measured in terms of the height of mercury in a barometer, is 4.58 mm at 0°C; and 760 mm at 100°C; (its boiling point).

THE BASIC DISTILLATION PROCESS

A simple distillation apparatus consists essentially of three parts: a flask equipped with a thermometer and with an outlet tube from which the vapour is emitted; a condenser that consists of two tubes of different diameters placed one within the other and so arranged that the smaller (in which the vapour is condensed) is held in a stream of coolant in the larger; and a vessel in which the condensed vapour is collected. The mixture of substances is placed in the flask and heated. Ideally, the substance with the lowest boiling point boiling point, temperature at which a substance changes its state from liquid to gas. A stricter definition of boiling point is the temperature at which the liquid and vapour (gas) phases of a substance can exist in equilibrium.vapourization, change of a liquid or solid substance to a gas or vapour.

There is fundamentally no difference between the terms *gas* and *vapour,* but gas is used commonly to describe a substance that appears in the gaseous state under standard conditions ofthe temperature remaining constant until that substance has completely distilled. The vapour is led into the condenser where, on being cooled, it reverts to the liquid (condenses) and runs off into a receiving vessel. The product so obtained is known as the distillate. Those substances having a higher boiling point remain in the flask and constitute the residue.Since a perfect separation is never effected, the distillate is often redistilled to increase its purity (hence the expression "double distilled" or "triple distilled"). Many alcoholic beverages are distilled, e.g., brandy, gin, whiskey, and various liqueurs. The apparatus used, called the still, is the same in principle as other distillation apparatus.

THE FRACTIONAL DISTILLATION PROCESS

When the substance with the lowest boiling point has been removed, the temperature can be raised and the distillation process repeated with the substance having the next lowest boiling point. The process of obtaining portions (or fractions) in this way is one type of fractional distillation. A more efficient method of fractional distillation involves placing a vertical tube called a fractionating column between the flask and the condenser. The column is filled with many objects on which the vapour can repeatedly condense and reevapourate as it moves toward the top, effectively distilling the vapour many times. The less volatile substances in the vapour tend to run back down the column after they condense, concentrating themselves near the bottom. The more volatile ones tend to reevapourate and keep moving upward, concentrating themselves near the top. Because of this the column can be tapped at various levels to draw off different fractions. Fractional distillation is commonly used in refining petroleum, some of the fractions thus obtained being gasoline, benzene, kerosene, fuel oils, lubricating oils, and paraffin.

THE DESTRUCTIVE DISTILLATION PROCESS

Another form of distillation involves heating out of free contact with air such substances as wood, coal, and oil shale and collecting separately the portions driven off; this is known as destructive distillation. Wood, when treated in this way yields acetic acid, methyl or wood alcohol, charcoal, and a number of hydrocarbons. Coal yields coal gas, coal tar, ammonia, and coke. Ammonia is also obtained by the destructive distillation of oil shale.

Petrochemical, any one of a large group of chemicals derived from a component of petroleum or natural gas. The cracking processes for manufacturing gasoline produce vast quantities of gaseous hydrocarbons. Originally considered waste products suitable only for use as illuminants and fuels, the gases today are manufactured into petrochemical substances widely employed in industry. Important petrochemical compounds are alcohols and aldehydes, butylene, butadiene, ethylene, propylene, toluene, styrene, acetylene, benzene, ethylene oxide, ethylene glycol, acrylonitrile, acetone, acetic acid, acetic anhydride, and ammonia. Materials made from the gases include carbon black, synthetic rubber, . polystyrene, polypropylene, and polyethylene. Petrochemicals are widely used in agriculture, in the manufacture of plastics, synthetic fibres, and explosives, and in the aircraft and automobile industries.

DEVELOPMENT OF PETROLEUM

Petroleum has been known throughout historical time. It was used in mortar, for coating walls and boat hulls, and as a fire weapon in defensive warfare. Native Americans used it in magic and medicine and in making paints. Pioneers bought it from the Native Americans for medicinal use and called it Seneca oil and Genesee oil. In Europe it was scooped from streams or holes in the ground, and in the early 19th cent. small quantities were made from shale. In 1815 several streets in Prague were lighted with petroleum lamps. Many wells were drilled in the region. Kerosene was the chief finished product, and kerosene lamps soon replaced whale oil lamps and candles in general use. Little use other than as lamp fuel was made of petroleum until the development of the gasoline engine and its application to automobiles, trucks, tractors, and airplanes. Today the world is heavily dependent on petroleum for motive power, lubrication, fuel, dyes, drugs, and many synthetics. The widespread use of petroleum has created serious environmental problems.

The great quantities that are burned as fuels generate most of the air pollution, contamination of the environment as a result of human activities. The term *pollution* refers primarily to the fouling of air, water, and land by wastes oil industry, the business of discovering oil (petroleum oily, flammable liquid that occurs naturally in deposits, usually beneath the surface of the earth; it is also called crude oil. It consists principally of a mixture of hydrocarbons, with traces of various nitrogenous and sulfurous compounds.extracting it from the ground, refining it into a variety of products, and distributing it to the public. The development of the oil industry in the 19th and 20th cent. provided a source of energy that now supplies about two fifths of the world's energy needs as well as a raw material that chemical and petroleum industries refine into a number of essential chemicals and industrial products.

Petroleum seeping out of underground reservoirs has been collected and used for light throughout recorded history. In the 4th cent. A.D. the Chinese drilled for oil and natural gas, but in the 1850s, oil was still being recovered by skimming it off the tops of ponds. As whale oil became less abundant, producers looked for new ways to extract oil. Edwin Drake dug the first modern oil well in Titusville, Pa, hitting oil at 69.5 ft (21.2 m), touching off an oil rush in the area. (Most modern wells go down over 4,700 ft (1,432 m).) In 1861 the first oil refinery was set up.

Late-Twentieth-Century and Early-Twenty-First-Century Developments

OPEC required that the major oil companies provide them with a larger percentage of the profits from their fields. After the oil embargo in early1970's.The resulting energy crisis forced many developing countries to pay more for energy, negatively affecting Third World debt; industrialized countries implemented new measures to conserve and develop new sources of energy. With an abundant supply, oil prices dropped and stayed low through the 1990s, until early1990's when OPEC announced that it would cut production in order to increase oil prices worldwide. With the help of non-OPEC oil-producing nations, and world events and demand have driven prices significantly higher.Economies dependent on oil production remain subject to the gyrations of the market. The collapse of oil prices in the mid-1980s ruined many independent refiners and helped produce a recession in such states as Texas. In contrast, the rise in oil prices early1990's has been responsible for economic growth in Russia, Venezuela, and other oil producers.

Improved recovery methods combined with higher prices that justify more expensive extraction costs have rejuvenated production in some older oil fields, increased the estimates of reserves in existing fields, and made feasible the exploitation of deposits once considered uneconomical.Many oil-producing nations in the Middle East and Latin America have set up their own refining operations since the 1970s, and state-owned oil companies in OPEC countries are now among the world's largest. Many large oil companies have diversified into chemicals, and oil prices are increasingly set on commodity trading exchanges such as the New York Mercantile Exchange. Beginning in the late 1990s, the industry saw increased consolidation as already large oil companies merged with each other, including Exxon with Mobil (the second largest; forming ExxonMobil), Chevron with Texaco and Unocal as Chevron, British Petroleum with Amoco and ARCO as BP, and Conoco with Phillips Petroleum as ConocoPhillips.

7

The Atmosphere

INTRODUCTION

The atmosphere is the gasesous envelope that surrounds the Earth and constitutes the transition between its surface and the vacuum of space. The atmosphere consists of a mixture of gases composed primarily of nitrogen, oxygen, carbon dioxide, and water vapour. It extends some 500 km above the surface of the Earth and the lower level (troposphere) constitutes the climate system that maintains the conditions suitable for life on the planet's surface. The next atmospheric level, the stratosphere (12 to 48 km), contains the ozone layer that protects life on the planet by filtering harmful ultraviolet radiation from the Sun. Since the Industrial Revolution, man has been altering the composition of the atmosphere by the burning of fossil fuels. Concern has been growing about rising concentrations of carbon dioxide, methane, nitrous oxide, and chloroflurocarbons in the atmosphere because these "greenhouse" gases trap heat energy emitted from the earth surface and increase global temperatures (global warming). In addition, chloroflurocarbons are effective at depleting the Earth's ozone shield that protects the earth surface from the harmful effects of ultraviolet radiation.A planet's climate is decided by its mass, its distance from the sun and the composition of its atmosphere. Mars is too small to keep a thick atmosphere. Its atmosphere consists mainly of carbon dioxide, but the atmosphere is very thin. The atmosphere of the Earth is a hundred times thicker. Most of Mars' carbon dioxide is frozen in the ground. Mars' average surface temperature is about -50°C. Venus has almost the same mass as Earth but a thicker atmosphere, composed of 96% carbon dioxide.

The surface temperature on Venus is +460°C. Earth's atmosphere is 78% nitrogen, 21% oxygen, and 1% other gases. Carbon dioxide accounts for just 0.03 - 0.04%. Water vapour, varying in amount from 0 to 2%, carbon dioxide and some other minor gases present in the atmosphere absorb some of the thermal radiation leaving the surface and emit radiation from much higher and colder levels out to space. These radiatively active gases are known as greenhouse gases because they act as a partial blanket for the thermal radiation from the surface and enable it to be substantially warmer than it would otherwise be, analogous to the effect of a greenhouse. This blanketing is known as the natural greenhous effect. Without the greenhouse gases, Earth's average temperature would be roughly -20°C. The climates on Mars and Venus are very different, but very stable and highly predictable. The Earth's climate is unstable and rather unpredictable as compared with that of the other two planets.

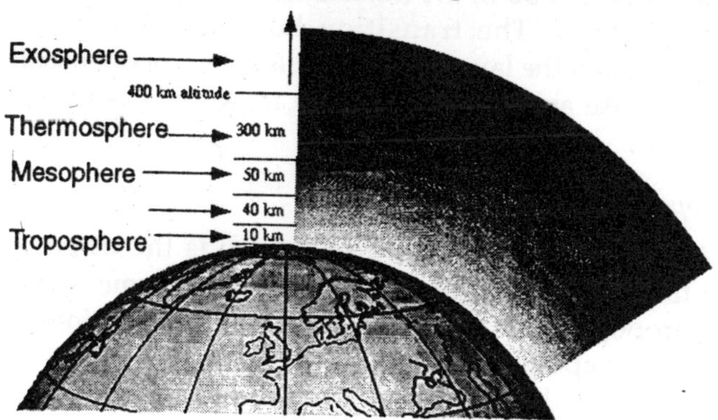

Fig:Layers of Atmosphere

The atmosphere is divided into five layers. It is thickest near the surface and thins out with height until it eventually merges with space.

- The troposphere is the first layer above the surface and contains half of the Earth's atmosphere. Weather occurs in this layer.
- Many jet aircrafts fly in the stratosphere because it is very stable. Also, the ozone layer absorbs harmful rays from the Sun.
- Meteors or rock fragments burn up in the mesosphere.
- The thermosphere is a layer with auroras. It is also where the space shuttle orbits.
- The atmosphere merges into space in the extremely thin exosphere. This is the upper limit of our atmosphere.

Troposphere

The troposphere begins at the Earth's surface and extends up to 4-12 miles (6-20 km) high. This is where we live. As the gases in this layer decrease with height, the air become thinner. Therefore, the temperature in the troposphere also decreases with height. As you climb higher, the temperature drops from about 62°F (17°C) to -60°F (-51°C). Almost all weather occurs in this region.

The height of the troposphere varies from the equator to the poles. At the equator it is around 11-12 miles (18-20 km) high, at 50°N and 50°S, 5½ miles and at the poles just under four miles high. The transition boundary between the troposphere and the layer above is called the tropopause. Both the tropopause and the troposphere are known as the lower atmosphere.

The Tropopause

At the very top of the troposphere is the tropopause where the temperature reaches a minimum. Some scientists call the tropopause a "cold trap" because this is a point where rising water vapour cannot go higher because it changes into ice and is trapped. If there is no cold trap, Earth would loose all its water!

The uneven heating of the regions of the troposphere by the Sun causes convection currents and winds. Warm air from Earth's surface rises and cold air above it rushes in to replace it. When warm air reaches the tropopause, it cannot go higher as the air above it (in the stratosphere) is warmer and lighter... preventing much air convection beyond the tropopause. The tropopause acts like an invisible barrier and is the reason why most clouds form and weather phenomena occur within the troposphere.

The Greenhouse Effect

Heat from the Sun warms the Earth's surface but most of it is radiated and sent back into space. Water vapour and carbon dioxide in the troposphere trap some of this heat, preventing it from escaping thus keep the Earth warm. This trapping of heat is called the "greenhouse effect".

However, if there is too much carbon dioxide in the troposphere then it will trap too much heat. Scientists are afraid that the increasing amounts of carbon dioxide would raise the Earth's surface temperature, bringing significant changes to worldwide weather patterns... shifting in climatic zones and the melting of the polar ice caps, which could raise the level of the world's oceans.

Stratosphere

The Stratosphere extends from the tropopause up to 31 miles above the Earth's surface. This layer holds 19 percent of the atmosphere's gases and but very little water vapour.

Temperature increases with height as radiation is increasingly absorbed by oxygen molecules which leads to the formation of Ozone. The temperature rises from an average -76°F (-60°C) at tropopause to a maximum of about 5°F (-15°C) at the stratopause due to this absorption of ultraviolet radiation. The increasing temperature also makes it a calm layer with movements of the gases slow.

The regions of the stratosphere and the mesosphere, along with the stratopause and mesopause, are called the middle atmosphere by scientists. The transition boundary which separates the stratosphere from the mesosphere is called the stratopause.

Mesosphere

The mesosphere extends from the stratopause to about 53 miles (85 km) above the earth. The gases, including the oxygen molecules, continue to become thinner and thinner with height. As such, the effect of the warming by ultraviolet radiation also becomes less and less leading to a decrease in temperature with height. On average, temperature decreases from about 5°F (-15°C) to as low as -184°F (-120°C) at the mesopause. However, the gases in the mesosphere are thick enough to slow down meteorites hurtling into the atmosphere, where they burn up, leaving fiery trails in the night sky.

Thermosphere

The Thermosphere extends from the mesopause to 430 miles (690 km) above the earth. This layer is known as the upper atmosphere.

The gases of the thermosphere are increasingly thinner than in the mesosphere. As such, only the higher energy ultraviolet and x-ray radiation from the sun is absorbed. But because of this absorption, the temperature increases with height and can reach as high as 3,600°F (2000°C) near the top of this layer.

However, despite the high temperature, this layer of the atmosphere would still feel very cold to our skin because of the extremely thin air. The total amount of energy from the very few molecules in this layer is not sufficient enough to heat our skin.

Exosphere

The Exosphere is the outermost layer of the atmosphere and extends from the thermopause to 6200 miles (10,000 km) above the earth. In this layer, atoms and molecules escape into space and satellites orbit the earth. The transition boundary which separates the exosphere from the thermosphere below it is called the thermopause.

THE OZONE LAYER

"The ozone layer" refers to the ozone within stratosphere, where over 90% of the earth's ozone resides. Ozone is an irritating, corrosive, colourless gas with a smell something like burning electrical wiring. In fact, ozone is easily produced by any high-voltage electrical arc (spark plugs, Van de Graaff generators, Tesla coils, arc welders). Each molecule of ozone has three oxygen atoms and is produced when oxygen molecules are broken up by energetic electrons or high energy radiation.

Ozone losses in the upper atmosphere are occurring at all latitudes in both hemispheres. The most striking example of ozone loss occurs over the South Pole during September and October. As ozone is lost, the amount of biologically harmful UV-B radiation will increase. Skin cancer rates are expected to increase. Other health effects will likely include an increase in cataracts and suppression of the immune system. Increased UV-B radiation may also harm plants and animals.

STRUCTURE OF THE ATMOSPHERE

The Atmosphere is divided into layers according to major changes in temperature. Gravity pushes the layers of air down on the earth's surface. This push is called air pressure. 99% of the total mass of the atmosphere is below 32 kilometers.

Troposphere

Troposphere - 0 to 12 km - Contains 75% of the gases in the atmosphere. This is where you live and where weather occurs. As height increases, temperature decreases. The temperature drops about 6.5 degrees Celsius for every kilometer above the earth's surface.

Tropopause

Tropopause - located at the top of the troposhere. The temperature remains fairly constant here. This layer separates the troposphere from the stratosphere. We find the jet stream here. These are very strong winds that blow eastward.

Stratosphere

Stratosphere - 12 to 50 km - in the lower part of the stratosphere. The temperature remains fairly constant (-60 degrees Celsius). This layer contains the ozone layer. Ozone acts as a shield for in the earth's surface. It absorbs ultraviolet radiation from the sun. This causes a temperature increase in the upper part of the layer.

Mesophere

Mesophere - 50 to 80 km - in the lower part of the stratosphere. The temperature drops in this layer to about -100 degrees Celsius. This is the coldest region of the atmosphere. This layer protects the earth from meteoroids. They burn up in this area.

Thermosphere

Thermosphere - 80 km and up - The air is very thin. Thermosphere means "heat sphere". The temperature is very high in this layer because ultraviolet radiation is turned into heat. Temperatures often reach 2000 degrees Celsius or more. This layer contains:

Ionosphere

Ionosphere - This is the lower part of the thermosphere. It extends from about 80 to 550 km. Gas particles absorb ultraviolet and X-ray radiation from the sun. The particles of gas become electrically charged (ions). Radio waves are bounced off the ions and reflect waves back to earth. This generally helps radio communication. However, solar flares can increase the number of ions and can interfere with the transmission of some radio waves.

Exosphere

Exosphere - the upper part of the thermosphere. It extends from about 550 km for thousands of kilometers. Air is very thin here. This is the area where satellites orbit the earth.

Magnetosphere

Magnetosphere - the area around the earth that extends beyond the atmosphere. The earth's magnetic field operates here. It begins at about 1000 km. It is made up of positively charged protons and negatively charged electrons. This traps the particles that are given off by the sun. They are concentrated into belts or layers called the Van Allen radiation belts. The Van Allen belts trap deadly radiation. When large amounts are given off during a solar flare, the particles collide with each other causing the aurora borealis or the northern lights.

LAYERS OF THE ATMOSPHERE

In a way, Earth's atmosphere is like an ocean of air that surrounds the land. It extends roughly 500 kilometers above the surface of the Earth and is split into layers. The layers are defined by sharp changes in temperature when moving vertically through the atmosphere. They are from the bottom of the atmosphere up to space: troposphere, stratosphere, mesosphere, thermosphere, and exosphere.

Layer of the Atmosphere	Relative Temperature	Characteristics of the Layer
Exosphere (500 km +)	very hot (very few air molecules)	-satellites orbit Earth in this layer -gradually blends with outer space
Thermosphere (85-500km)	very hot (very few air molecules)	-the ionosphere is in this layer (causes the northern lights)
Mesosphere (50-85 km)	very cold (coldest layer)	-meteors often burn up in this layer due to friction with air molecules
Stratosphere (10-50 km)	warms up with height since the absorption of ozone causes this layer to warm up	-some weather balloons fly up to this layer -the jet stream is located right at the lower boundary of this layer
Troposphere (0-10 km)	gets colder with height	-75% of all the atmospheric gases are in this layer -all weather events are

in this layer-the top of Mt. Everest just barely reaches the top of this layer

GASES IN THE ATMOSPHERE

Earth's atmosphere is made of a mixture of gases. Surprisingly, oxygen is not the most abundant gas in the atmosphere. That honour belongs to nitrogen at roughly 78% of all the gases. Oxygen comes in second at around 21%. The last 1% or so of the atmosphere is made up of such small amounts of other gases that it doesn't make sense to list each gas separately, so they are lumped together and called trace gases. Water vapour is another gas in Earth's atmosphere and it ranges from 0%-4% of the gases in the atmosphere, depending on temperature. The other gases in the atmosphere adjust to fit the water vapour.

Ozone is another important gas in the atmosphere. The stratosphere has the highest amount of naturally occurring ozone. Ozone in the troposphere is considered a pollutant. Ozone is made of three atoms of oxygen bonded together to make one molecule of ozone. Ozone is destroyed by chemicals caused by chloroflurocarbons (CFCs). Researchers think that chlorine from these types of chemicals breaks the bonds between the oxygen atoms. This produces one molecule of regular oxygen, and one molecule of chlorine and oxygen bonded together. Along comes a free floating oxygen that bumps out the chlorine and bonds with the other oxygen. Now there is another regular molecule of oxygen and a free floating atom of chlorine. This chlorine is now ready to go and destroy another molecule of ozone. This is bad for the ozone layer. Ozone absorbs ultraviolet radiation from the sun. Without it, skin cancer rates would increase.

PRESSURE IN THE ATMOSPHERE

In general, as one moves away from the surface of the Earth, the air pressure lessens. As one moves higher into the atmosphere, the amount of air pressing down on an object is less than it would be if it were at the surface. Kind of like a football tackle. The player at the bottom of the pile feels the most pressure because of all the other players piled up on top. The player on the top of the pile feels the least pressure since there aren't any players higher up. Changes in pressure can occur because of changes in temperature. This also results in changes in density of the gas. Warm air is less dense than cold air, therefore it rises and creates areas of low pressure. Cold air is more dense than warm air, therefore it sinks and creates areas of high pressure. This is a very important concept to understand in weather.

ENERGY TRANSFER IN THE ATMOSPHERE

The energy that heats Earth's atmosphere comes from the sun in the form of radiation. Radiation is the transfer of energy in the form of electromagnetic waves. These waves travel from the sun, through space, down through Earth's atmosphere, where they hit the surface of Earth. The surfaces on Earth absorb the radiation and become warm. These warm surfaces are in direct contact with the air above the ground. The air above the ground becomes warm because the ground conducts the heat from the ground to the air. Conduction is the transfer of energy due to direct contact.

This warm air is less dense than the cooler air surrounding it, and therefore will rise higher in the troposphere. This is an example of convection. Convection is the transfer of energy due to the flow of a heated material. Using radiation, conduction and convection, the lower portion of the troposphere is heated. The temperature difference between the warm and cold air creates a difference in density; warm air is less dense than cold air. As the warm air rises, it creates a low pressure area on the surface of Earth. As cold air sinks, it creates a high pressure area on the surface of Earth. When convection occurs, winds, storms, and clouds form.

Type of Energy Transfer	How Energy is Transferred
Radiation	Electromagnetic waves
Conduction	Direct Contact
Convection	Differences in Density

WIND-FORMATION

Wind is caused by the uneven heating and cooling of the Earth. As warm air rises, it creates a low pressure area on the surface of Earth. As cold air sinks, it creates a high pressure area on the surface of Earth. As the warm air rises it pushes cooler air in to take the warm air's place. This circular movement of air is called a convection current. The horizontal movement of air on the surface of the Earth is known as wind.

Wind-local Winds

Local winds are directly related to convection currents. These local winds blow around in a rather small scale (on the order of a town or so). In order to have wind, differences in pressure caused by differences in density must be present. These differences are caused by the uneven heating and cooling rates of land and water. Land both heats and cools more quickly than water does. That means that in the day, the land will heat up rather quickly, but a water source such as a lake will take considerably more time to heat up. On the other hand, once water is heated, it will take a relatively long time for it to cool down. You might have noticed this in the early summer (like in June) when you want to go swimming in a lake. Walking across the parking lot or sand, your bare feet feel strong heat, but the water itself will feel pretty cold. If you wait until evening, the water will feel warmer than it did during the day because it has now had all day to heat up. The sand or parking lot will feel cold since most of the heat has dissipated.

This uneven heating means that air over land during the day will be warmer than the air over water during the day. Warm air over the land will be rising, where it will cool and sink back to the Earth over the water. Air from over the water will rush in to take the place of the rising warm air over the land. This breeze is called a sea breeze (or lake breeze) since it is coming from over the water. At night, the air over the water is warmer than the air over the land (since it has had all day to heat up, and it takes water longer to cool down than land does) so the air over the water rises, where it cools and sinks back down over the land. This air rushes in to take the place of the air rising over the water. This breeze is called a land breeze since it is coming from the land.

Wind-Global Winds

Global winds are also caused by convection. These are belts of wind that circle the globe. A couple of things to think about when talking about global wind belts.:

- Air would flow from the pole to the equator in a straight line, if it could (because cold air is more dense than warm air so it sinks at the poles, and rushes along the surface of the Earth towards the equator).
- Air doesn't flow in that straight path because of Earth's rotation. Earth's rotation causes the Coriolis Effect. The Coriolis Effect is the deflection of wind to the right (northern hemisphere) or left (southern hemisphere).
- If the Earth had no land masses, the winds would spiral around the globe in both hemispheres. Since the Earth has land masses, however, this swirling pattern is disrupted.
- This disruption causes the winds to be organized into belts or zones of wind. Moving either north or south of the equator they are: trade winds, prevailing westerlies and polar easterlies. These winds are named for the direction they come from. Right around the equator (from about 10° N latitude to about 10° S latitude) are the doldrums, or windless zone. Being a sailor stuck in the doldrums would be a very bad idea!

Colour of the Sky

It is easy to see that the sky is blue. The light from the Sun looks white. But it is really made up of all the colours of the rainbow.

A prism is a specially shaped crystal. When white light shines through a prism, the light is separated into all its colours.

Like energy passing through the ocean, light energy travels in waves, too. Some light travels in short, "choppy" waves. Other light travels in long, lazy waves. Blue light waves are shorter than red light waves.

All light travels in a straight line unless something gets in the way to—

- Reflect it (like a mirror)
- Bend it (like a prism)
- or scatter it (like *molecules* of the gases in the atmosphere)

Sunlight reaches Earth's atmosphere and is scattered in all directions by all the gases and particles in the air. Blue light is scattered in all directions by the tiny molecules of air in Earth's atmosphere. Blue is scattered more than other colours because it travels as shorter, smaller waves. This is why we see a blue sky most of the time.

Closer to the horizon, the sky fades to a lighter blue or white. The sunlight reaching us from low in the sky has passed through even more air than the sunlight reaching us from overhead. As the sunlight has passed through all this air, the air molecules have scattered and rescattered the blue light many times in many directions. Also, the surface of Earth has reflected and scattered the light. All this scattering mixes the colours together again white and less blue.

Tropospheric Ozone, the Polluter

Ozone occurs naturally at ground-level in low concentrations. The two major sources of natural ground-level ozone are hydrocarbons, which are released by plants and soil, and small amounts of stratospheric ozone, which occasionally migrate down to the earth's surface.

Fig.: Comet

Neither of these sources contributes enough ozone to be considered a threat to the health of humans or the environment.

But the ozone that is a byproduct of certain human activities does become a problem at ground level and this is what we think of as 'bad' ozone. With increasing populations, more automobiles, and more industry, there's more ozone in the lower atmosphere. Since 1900 the amount of ozone near the earth's surface has more than doubled. Unlike most other air pollutants, ozone is not directly emitted from any one source. Tropospheric ozone is formed by the interaction of sunlight, particularly ultraviolet light, with hydrocarbons and nitrogen oxides, which are emitted by automobiles, gasoline vapours, fossil fuel power plants, refineries, and certain other industries.

High ozone levels usually occur during the warm, sunny summer months (from May through September). Typically, ozone levels reach their peak in mid to late afternoon, after the sun has had time to react fully with the exhaust fumes from the morning rush hours. A hot, sunny, still day is the perfect environment for ozone pollution production. In early evening, the sunlight's intensity decreases and the photochemical production process that forms ground level ozone begins to subside.

Negative Impacts of Tropospheric Ozone

While stratospheric ozone shields us from ultraviolet radiation, in the troposphere this irritating, reactive molecule damages forests and crops; destroys nylon, rubber, and other materials; and injures or destroys living tissue. It is a particular threat to people who exercise outdoors or who already have respiratory problems.

Ozone affects plants in several ways. High concentrations of ozone cause plants to close their stomata. These are the cells on the underside of the plant that allow carbon dioxide and water to diffuse into the plant tissue. This slows down photosynthesis and plant growth. Ozone may also enter the plants through the stomata and directly damage internal cells.

Fi.g.: Ozone injury to yellow-popular Ozone injury to milkweed

Rubber, textile dyes, fibres, and certain paints may be weakened or damaged by exposure to ozone. Some elastic materials can become brittle and crack, while paints and fabric dyes may fade more quickly.

When ozone pollution reaches high levels, pollution alerts are issued urging people with respiratory problems to take extra precautions or to remain indoors. Smog can damage respiratory tissues through inhalation. Ozone has been linked to tissue decay, the promotion of scar tissue formation, and cell damage by oxidation. It can impair an athlete's performance, create more frequent attacks for individuals with asthma, cause eye irritation, chest pain, coughing, nausea, headaches and chest congestion and discomfort. It can worsen heart disease, bronchitis, and emphysema.

So why can't we take all of this "bad" ozone and blast it up into the stratosphere. The answer lies in the vast quantities needed and ozone's instability in the dynamic atmosphere. Ozone molecules don't last very long, with or without human intervention. The vehicle necessary to transport such enormous amounts of ozone into the stratosphere does not exist, and, if it did, it would require so much fuel that the resulting pollution might undo any positive effect. Rather than seek such grandiose solutions, we need to decrease the production of those chemicals that break down ozone in the stratosphere and help create ozone in the troposphere.

The dual ozone problems—pollution or smog in the troposphere and depletion of the ozone layer in the stratosphere—are indeed very different. But the problems have common ties in that they both are related to air pollutants that come from industry, transportation, and other human activities.

Concluding Thoughts

Citizens live in areas that are impacted by tropospheric ozone pollution. They are familiar with "smog-alerts," local government pleas to reduce vehicle traffic, and news reports about cities that have failed to meet EPA standards for ozone pollution levels. As you work through these activities with your class, you can easily connect the instructional activities to your students' surroundings by collecting and discussing news reports about smog issues in your own or nearby metropolitan areas. You may further wish to engage students in a discussion (or perhaps even a research project) about the impact of private automobiles on ozone levels and the prospects for alternative-fuel vehicles to reduce vehicle emissions and tropospheric ozone.

Stratospheric Ozone, the Protector

The debate over the existence of an ozone problem breeds media coverage. However, the real story is not whether stratospheric ozone levels are decreasing, but what those decreases may mean for life on earth. As the percentage of ozone in the atmosphere decreases, the amount of UV-B radiation reaching the surface increases. It's the UV-B radiation, not the ozone itself that concerns scientists, because the invisible wavelengths are linked to skin cancers and other biological damage.

Measuring UV-B is tricky. Levels are affected by time of day, day of the year, latitude, weather conditions, and the amount of ozone aloft. UV is the part of the electromagnetic spectrum made up of wavelengths between 280 and 400 nanometers (billionths of a meter). Most of this is UV-A light, only mildly associated with sunburn and DNA damage and relatively benign to most plant life. But the ill effects increase more than a thousandfold in the shorter wavelengths referred to as UV-B. Below 300 nanometers, the rays are sparse but very damaging; near 315 nanometers they're more numerous but much less destructive. Close to 310 nanometers lies the middle ground, where the number and impact of rays combine to cause the greatest harm to humans and plants. Engineers face enormous challenges when designing instruments that can measure individual wavelengths, yet such precision is necessary to determine the amount of dangerous light entering the atmosphere.

The Story of the Ozone Hole

Although often referred to as the ozone 'hole', it is really not a hole but rather a thinning of the ozone layer in the stratosphere. We will use the term 'hole' in reference to the seasonal thinning of the ozone layer.

The appearance of a hole in the earth's ozone layer over Antarctica, first detected in 1976, was so unexpected that scientists didn't pay attention to what their instruments were telling them; they thought their instruments were malfunctioning. When that explanation proved to be erroneous, they decided they were simply recording natural variations in the amount of ozone. It wasn't until 1985 that scientists were certain they were seeing a major problem.

Why did it take scientists so long to solve this mystery? To begin with, observations that challenge preconceived ideas don't always get taken seriously, even in science. Two decades ago scientists did not suspect the importance of the chemical processes that rapidly destroy ozone in the Antarctic stratosphere. When they saw dramatic fluctuations in ozone levels, they assumed their instruments were in error, or that whatever was happening was due to natural processes like sunspot activity or volcanic eruptions.

They didn't realise that chlorine was the main culprit and that most of the chlorine in the stratosphere comes from human activity. The largest source is a class of chemical compounds known as chlorofluorocarbons (CFCs).

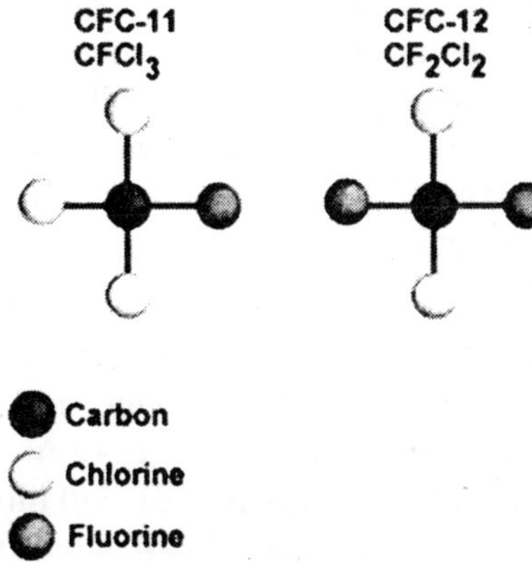

Fig.: Chlorofluorocarbons

Because of their chemical stability, low toxicity, and valuable physical properties, these chemicals, versatile and stable in the lower atmosphere, at least, have been extensively used since the 1960s as refrigerants, industrial cleaning solvents, propellants in aerosol spray cans, and to make Styrofoam.

Global Warming

Global Warming is caused by many things. The causes are split up into two groups, man-made or anthropogenic causes, and natural causes.

Natural Causes

Natural causes are causes created by nature. One natural cause is a release of methane gas from arctic tundra and wetlands. Methane is a greenhouse gas. A greenhouse gas is a gas that traps heat in the earth's atmosphere. Another natural cause is that the earth goes through a cycle of climate change. This climate change usually lasts about 40,000 years.

Man-made Causes

Man-made causes probably do the most damage. There are many man-made causes. Pollution is one of the biggest man-made problems. Pollution comes in many shapes and sizes. Burning fossil fuels is one thing that causes pollution.

Fossil fuels are fuels made of organic matter such as coal, or oil. When fossil fuels are burned they give off a green house gas called CO2. Also mining coal and oil allows methane to escape. How does it escape? Methane is naturally in the ground. When coal or oil is mined you have to dig up the earth a little. When you dig up the fossil fuels you dig up the methane as well.

Another major man-made cause of Global Warming is population. More people means more food, and more methods of transportation, right? That means more methane because there will be more burning of fossil fuels, and more agriculture. Now your probably thinking, "Wait a minute, you said agriculture is going to be damaged by Global Warming, but now you're saying agriculture is going to help cause Global Warming?" Well, have you ever been in a barn filled with animals and you smell something terrible? You're smelling methane. Another source of methane is manure. Because more food is needed we have to raise food. Animals like cows are a source of food which means more manure and methane. Another problem with the increasing population is transportation. More people means more cars, and more cars means more pollution. Also, many people have more than one car.

Since CO_2 contributes to global warming, the increase in population makes the problem worse because we breathe out CO_2. Also, the trees that convert our CO_2 to oxygen are being demolished because we're using the land that we cut the trees down from as property for our homes and buildings. We are not replacing the trees (an important part of our eco system), so we are constantly taking advantage of our natural resources and giving nothing back in return.

EFFECT FOR GLOBAL WARMING

The greenhouse effect

When sunlight reaches Earth's surface some is absorbed and warms the earth and most of the rest is radiated back to the atmosphere at a longer wavelength than the sun light.

Some of these longer wavelengths are absorbed by greenhouse gases in the atmosphere before they are lost to space. The absorption of this longwave radiant energy warms the atmosphere. These greenhouse gases act like a mirror and reflect back to the Earth some of the heat energy which would otherwise be lost to space. The reflecting back of heat energy by the atmosphere is called the "greenhouse effect".

The major natural greenhouse gases are water vapour, which causes about 36-70% of the greenhouse effect on Earth (not including clouds); carbon dioxide CO_2, which causes 9-26%; methane, which causes 4-9%, and ozone, which causes 3-7%. It is not possible to state that a certain gas causes a certain percentage of the greenhouse effect, because the influences of the various gases are not additive. Other greenhouse gases include, but are not limited to, nitrous oxide, sulfur hexafluoride, hydrofluorocarbons, perfluorocarbons and chlorofluorocarbons.

Global Warming causes by Greenhouse Effect

Greenhouse gases in the atmosphere act like a mirror and reflect back to the Earth a part of the heat radiation, which would otherwise be lost to space. The higher the concentration of green house gases like carbon dioxide in the atmosphere, the more heat energy is being reflected back to the Earth. The emission of carbon dioxide into the environment mainly from burning of fossil fuels (oil, gas, petrol, kerosene, etc.)

Effects of global warming

There are two major effects of global warming:

- Increase of temperature on the earth by about 3° to 5° C (34° to 41° Fahrenheit) by the year 2100.
- Rise of sea levels by at least 25 meters (82 feet) by the year 2100.

More details about the effects of global warming:

Increasing global temperatures are causing a broad range of changes. Sea levels are rising due to thermal expansion of the ocean, in addition to melting of land ice. Amounts and patterns of precipitation are changing. The total annual power of hurricanes has already increased markedly since 1975 because their average intensity and average duration have increased (in addition, there has been a high correlation of hurricane power with tropical sea-surface temperature).

Changes in temperature and precipitation patterns increase the frequency, duration, and intensity of other extreme weather events, such as floods, droughts, heat waves, and tornadoes. Other effects of global warming include higher or lower agricultural yields, further glacial retreat, reduced summer stream flows, species extinctions. As a further effect of global warming, diseases like malaria are returning into areas where they have been extinguished earlier.

Although global warming is affecting the number and magnitude of these events, it is difficult to connect specific events to global warming. Although most studies focus on the period up to 2100, warming is expected to continue past then because carbon dioxide (chemical symbol CO_2) has an estimated atmospheric lifetime of 50 to 200 years. For a summary of the predictions for the future increase in temperature up to 2100.

Atmospheric Composition

Nitrogen and oxygen are the main components of the atmosphere by volume. Together these two gases make up approximately 99% of the dry atmosphere. Both of these gases have very important associations with life. Nitrogen is removed from the atmosphere and deposited at the Earth's surface mainly by specialized nitrogen fixing bacteria, and by way of lightning through precipitation. The addition of this nitrogen to the Earth's surface soils and various water bodies supplies much needed nutrition for plant growth. Nitrogen returns to the atmosphere primarily through biomass combustion and denitrification.

Oxygen is exchanged between the atmosphere and life through the processes of photosynthesis and respiration. Photosynthesis produces oxygen when carbon dioxide and water are chemically converted into glucose with the help of sunlight. Respiration is a the opposite process of photosynthesis. In respiration, oxygen is combined with glucose to chemically release energy for metabolism. The products of this reaction are water and carbon dioxide.

The next most abundant gas on the table is water vapour. Water vapour varies in concentration in the atmosphere both spatially and temporally. The highest concentrations of water vapour are found near the equator over the oceans and tropical rain forests. Cold polar areas and subtropical continental deserts are locations where the volume of water vapour can approach zero percent. Water vapour has several very important functional roles on our planet:

- It redistributes heat energy on the Earth through latent heat energy exchange.
- The condensation of water vapour creates precipitaion that falls to the Earth's surface providing needed fresh water for plants and animals.
- It helps warm the Earth's atmosphere through the greenhouse effect.

The fifth most abundant gas in the atmosphere is carbon dioxide. The volume of this gas has increased by over 35% in the last three hundred years. This increase is primarily due to human induced burning from fossil fuels, deforestation, and other forms of land-use change. Carbon dioxide is an important greenhouse gas. The human-caused increase in its concentration in the atmosphere has strengthened the greenhouse effect and has definitely contributed to global warming over the last 100 years. Carbon dioxide is also naturally exchanged between the atmosphere and life through the processes of photosynthesis and respiration.

Methane is a very strong greenhouse gas.Methane concentrations in the atmosphere have increased by more than 150%. The primary sources for the additional methane added to the atmosphere (in order of importance) are: rice cultivation; domestic grazing animals; termites; landfills; coal mining; and, oil and gas extraction. Anaerobic conditions associated with rice paddy flooding results in the formation of methane gas. However, an accurate estimate of how much methane is being produced from rice paddies has been difficult to ascertain. More than 60% of all rice paddies are found in India and China where scientific data concerning emission rates are unavailable. Nevertheless, scientists believe that the contribution of rice paddies is large because this form of crop production has more than doubled since 1950. Grazing animals release methane to the environment as a result of herbaceous digestion. Some researchers believe the addition of methane from this source has more than quadrupled over the last century.

Termites also release methane through similar processes. Land-use change in the tropics, due to deforestation, ranching, and farming, may be causing termite numbers to expand. If this assumption is correct, the contribution from these insects may be important. Methane is also released from landfills, coal mines, and gas and oil drilling. Landfills produce methane as organic wastes decompose over time. Coal, oil, and natural gas deposits release methane to the atmosphere when these deposits are excavated or drilled.

The average concentration of the greenhouse gas nitrous oxide is now increasing at a rate of 0.2 to 0.3% per year. Its part in the enhancement of the greenhouse effect is minor relative to the other greenhouse gases. However, it does have an important role in the artificial fertilization of ecosystems. In extreme cases, this fertilization can lead to the death of forests, eutrophication of aquatic habitats, and species exclusion. Sources for the increase of nitrous oxide in the atmosphere include: land-use conversion; fossil fuel combustion; biomass burning; and soil fertilization.

Most of the nitrous oxide added to the atmosphere each year comes from deforestation and the conversion of forest, savanna and grassland ecosystems into agricultural fields and rangeland. Both of these processes reduce the amount of nitrogen stored in living vegetation and soil through the decomposition of organic matter. Nitrous oxide is also released into the atmosphere when fossil fuels and biomass are burned. However, the combined contribution to the increase of this gas in the atmosphere is thought to be minor. The use of nitrate and ammonium fertilizers to enhance plant growth is another source of nitrous oxide. How much is released from this process has been difficult to quantify. Estimates suggest that the contribution from this source represents from 50% to 0.2% of nitrous oxide added to the atmosphere annually.

Ozone's role in the enhancement of the greenhouse effect has been difficult to determine. Accurate measurements of past long-term (more than 25 years in the past) levels of this gas in the atmosphere are currently unavailable. Moreover, concentrations of ozone gas are found in two different regions of the Earth's atmosphere. The majority of the ozone (about 97%) found in the atmosphere is concentrated in the stratosphere at an altitude of 15 to 55 kilometers above the Earth's surface. This stratospheric ozone provides an important service to life on the Earth as it absorbs harmful ultraviolet radiation. In recent years, levels of stratospheric ozone have been decreasing due to the buildup of human created chlorofluorocarbons in the atmosphere. Since the late 1970s, scientists have noticed the development of severe holes in the ozone layer over Antarctica. Satellite measurements have indicated that the zone from 65° North to 65° South latitude has had a 3% decrease in stratospheric ozone since 1978.

Ozone is also highly concentrated at the Earth's surface in and around cities. Most of this ozone is created as a by product of human created photochemical smog. This buildup of ozone is toxic to organisms living at the Earth's surface.

Atmospheric Effects on Incoming Solar Radiation

Three atmospheric processes modify the solar radiation passing through our atmosphere destined to the Earth's surface. These processes act on the radiation when it interacts with gases and suspended particles found in the atmosphere. The process of scattering occurs when small particles and gas molecules diffuse part of the incoming solar radiation in random directions without any alteration to the wavelength of the electromagnetic energy. Scattering does, however, reduce the amount of incoming radiation reaching the Earth's surface. A significant proportion of scattered shortwave solar radiation is redirected back to space. The amount of scattering that takes place is dependent on two factors: wavelength of the incoming radiation and the size of the scattering particle or gas molecule. In the Earth's atmosphere, the presence of a large number of particles with a size of about 0.5 microns results in shorter wavelengths being preferentially scattered. This factor also causes our sky to look blue because this colour corresponds to those wavelengths that are best diffused. If scattering did not occur in our atmosphere the daylight sky would be black.

The Greenhouse Effect

The greenhouse effect is a naturally occurring process that aids in heating the Earth's surface and atmosphere. It results from the fact that certain atmospheric gases, such as carbon dioxide, *water vapour*, and methane, are able to change the energy balance of the planet by absorbing longwave radiation emitted from the Earth's surface. Without the greenhouse effect life on this planet would probably not exist as the average temperature of the Earth would be a chilly -18° Celsius, rather than the present 15° Celsius.

As energy from the Sun passes through the atmosphere a number of things take place. A portion of the energy (26% globally) is reflected or scattered back to space by clouds and other atmospheric particles. About 19% of the energy available is absorbed by clouds, gases (like ozone), and particles in the atmosphere. Of the remaining 55% of the solar energy passing through the Earth's atmosphere, 4% is reflected from the surface back to space. On average, about 51% of the Sun's radiation reaches the surface. This energy is then used in a number of processes, including the heating of the ground surface; the melting of ice and snow and the evapouration of water; and plant photosynthesis.

8

Waste Materials Management

INTRODUCTION

The Waste and Materials Management Progammeme encourages management of waste as a resource to help ensure a clean and healthy Wisconsin for future generations.

Residential Recycling

We develop policies and offer technical assistance to actively encourage the reduction, recycling and re-use of wastes as raw material for new products. We oversee management of solid and hazardous waste through storage, treatment and disposal. We also work with local governments to reclaim mining sites to valued natural resources.

Our goal is to increase waste material reuse and recycling in Wisconsin by 30%. Wisconsin's communities and businesses benefit from a more efficient economy and a cleaner environment when waste becomes a resource.

Reducing Fugitive Landfill Gas Emissions

This innovative initiative, which involves no new regulations and relies heavily on voluntary actions by landfill owners, seeks to reduce fugitive landfill gas emissions. Information describing the initiative, graphs for individual municipal solid waste landfills

Burn Barrels: Unhealthy, Unneighbourly and Often Illegal

Backyard trash and leaf burning often releases high levels of toxic compounds, often including dioxin, furans and other carcinogens.

Hazardous Waste Regulatory Progammeme

The Wisconsin Department of Natural Resources operates a Federally authorized hazardous waste regulatory progammeme. Staff in this progammeme help to ensure the protection of Wisconsin's ground and surface water, air and soil by working with waste generators and facilities to encourage the proper management of hazardous wastes.In Wisconsin, there are over 11,000 businesses, schools and government institutions that generate varying quantities of hazardous wastes each year. The number of hazardous waste generators and the quantity of hazardous waste that they generate is declining each year as everyone learns how much it costs to generate wastes and manage hazardous wastes according to the strict requirements that apply. While much of the solvent-type hazardous wastes that are generated in Wisconsin are recycled here, many other hazardous wastes are handled out of state.Wisconsin also is home to 17 licensed hazardous waste management facilities. Many of these facilities are privately operated, serving the needs of that particular facility's hazardous wastes.

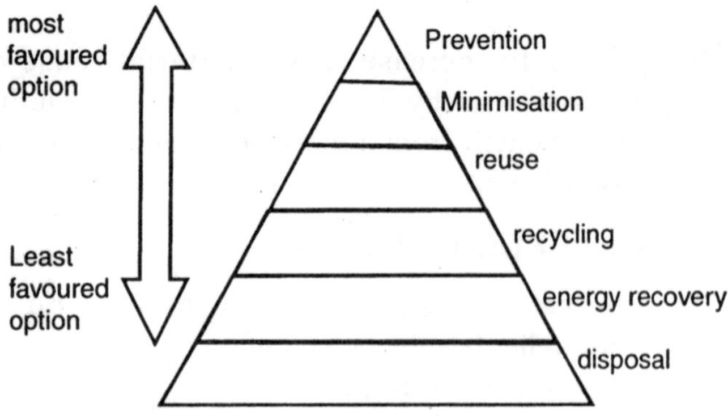

Fig.: Waste Management of Model

The commercial hazardous waste facilities in Wisconsin

primarily focus on recycling of hazardous waste solvents and mercury, fuel blending of hazardous wastes for energy recovery, storage of hazardous wastes prior to the treatment at licensed hazardous waste facilities in other states and treatment of hazardous wastes to facilitate disposal. There are no operating hazardous waste disposal facilities (i.e., landfills) in Wisconsin.

Waste management is the collection, transport, processing (waste treatment), recycling or disposal of waste materials, usually ones produced by human activity, in an effort to reduce their effect on human health or local aesthetics or amenity. A sub-focus in recent decades has been to reduce waste materials' effect on the natural world and the environment and to recover resources from them.Waste management can involve solid, liquid or gaseous substances with different methods and fields of expertise for each.Waste management practices differ for developed and developing nations, for urban and rural areas, and for residential, industrial, and commercial producers. Waste management for non-hazardous residential and institutional waste in metropolitan areas is usually the responsibility of local government authorities, while management for non-hazardous commercial and industrial waste is usually the responsibility of the generator.

Waste Management Concepts

There are a number of concepts about waste management, which vary in their usage between countries or regions.

The waste hierarchy:

- Reduce
- Reuse
- Recycle

Classifies waste management strategies according to their desirability. The waste hierarchy has taken many forms over the past decade, but the basic concept has remained the cornerstone of most waste minimization strategies. The aim of the waste hierarchy is to extract the maximum practical benefits from products and to generate the minimum amount of waste.

Some waste management experts have recently incorporated a 'fourth R': "Re-think", with the implied meaning that the present system may have fundamental flaws, and that a thoroughly effective system of waste management may need an entirely new way of looking at waste. Some "re-think" solutions may be counter-intuitive, such as cutting fabric patterns with slightly more "waste material" left — the now larger scraps are then used for cutting small parts of the pattern, resulting in a decrease in net waste. This type of solution is by no means limited to the clothing industry.

Source reduction involves efforts to reduce hazardous waste and other materials by modifying industrial production. Source reduction methods involve changes in manufacturing technology, raw material inputs, and product formulation. At times, the term "pollution prevention" may refer to source reduction.

Another method of source reduction is to increase incentives for recycling. Many communities in the United States are implementing variable rate pricing for waste disposal which has been effective in reducing the size of the municipal waste stream.Source reduction is typically measured by efficiencies and cutbacks in waste. Toxics use reduction is a more controversial approach to source reduction that targets and measures reductions in the upstream use of toxic materials. Toxics use reduction emphasis's the more preventive aspects of source reduction but, due to its emphasis on toxic chemical inputs, has been opposed more vigourously by chemical manufacturers. Toxics use reduction progammemes have been set up by legislation in some states.

Resource Recovery

A relatively recent idea in waste management has been to treat the waste material as a resource to be exploited, instead of simply a challenge to be managed and disposed of. There are a number of different methods by which resources may be extracted from waste: the materials may be extracted and recycled, or the calorific content of the waste may be converted to electricity.

The process of extracting resources or value from waste is variously referred to as secondary resource recovery, recycling, and other terms. The practice of treating waste materials as a resource is becoming more common, especially in metropolitan areas where space for new landfills is becoming scarcer. There is also a growing acknowledgement that simply disposing of waste materials is unsustainable in the long term, as there is a finite supply of most raw materials.

There are a number of methods of recovering resources from waste materials, with new technologies and methods being developed continuously.In some developing nations some resource recovery already takes place by way of manual labourers who sift through un-segregated waste to salvage material that can be sold in the recycling market. These unrecognized workers called waste pickers or rag pickers, are part of the informal sector, but play a significant role in reducing the load on the Municipalities' Solid Waste Management departments. There is an increasing trend in recognizing their contribution to the environment and there are efforts to try and integrate them into the formal waste management systems, which is proven to be both cost effective and also appears to help in urban poverty alleviation. However, the very high human cost of these activities including disease, injury and reduced life expectancy through contact with toxic or infectious materials would not be tolerated in a developed country.

Recycling

Recycling means to recover for other use a material that would otherwise be considered waste. The popular meaning of 'recycling' in most developed countries has come to refer to the widespread collection and reuse of various everyday waste materials. They are collected and sorted into common groups, so that the raw materials from these items can be used again (recycled).

In developed countries, the most common consumer items recycled include aluminum beverage cans, steel, food and aerosol cans, HDPE and PET plastic bottles, glass bottles and jars, paperboard cartons, newspapers, magazines, and cardboard. Other types of plastic are also recyclable, although not as commonly collected. These items are usually composed of a single type of material, making them relatively easy to recycle into new products.

The recycling of obsolete computers and electronic equipment is important, but more costly due to the separation and extraction problems. Much electronic waste is sent to Asia, where recovery of the gold and copper can cause environmental problems (monitors contain lead and various "heavy metals", such as selenium and cadmium; both are commonly found in electronic items).

Recycled or used materials have to compete in the marketplace with new (virgin) materials. The cost of collecting and sorting the materials often means that they are equally or more expensive than virgin materials. This is most often the case in developed countries where industries producing the raw materials are well-established. Practices such as trash picking can reduce this value further, as choice items are removed (such as aluminum cans). In some countries, recycling progammemes are subsidized by deposits paid on beverage containers The economics of recycling junked automobiles also depends on the scrap metal market except where recycling is mandated by legislation.However, most economic systems do not account for the benefits to the environment of recycling these materials, compared with extracting virgin materials. It usually requires significantly less energy, water and other resources to recycle materials than to produce new materials

- Recycling 1000 kg of aluminum cans saves approximately 5000 kg of bauxite ore being mined and prevents the generation of 15.17 tons CO_2 eq greenhouse gases
- Recycling steel saves about 95% of the energy used to refine virgin ore

CONSUMER Vs. MACHINE WASTE SEPARATION:

In many areas, material for recycling is collected separately from general waste, with dedicated bins and collection vehicles. Other waste management processes recover these materials from general waste streams. This usually results in greater levels of recovery than separate collections of consumer-separated beverage containers, but are more complex and expensive.

When consumer-separated recycling is a government requirement, waste is often not well separated due of either ignorance or contempt of the rules. This results in glass containers that may have metal lids still attached and rotted food inside, aluminum cans full of chewing tobacco spit and cigarette butts, corrugated paper boxes soiled with oils, solvents, or rotting food, and inclusion of incompatible plastic types in a plastics recycling bin. This can all lead to process contamination, work stoppage, a system cleanout, and landfill disposal of the contaminated batch of otherwise recyclable materials. Re-sorting of consumer-separated wastes is often needed to prevent recycling process contamination.A common method of machine sorting of complex waste streams is to shred the entire stream into a fine particulate of similar size. A magnetic conveyor belt removes ferrous metals from this particulate, and cyclonic separation towers separate objects from the waste stream by mass. Spectral imaging such as with X-rays can further separate glass and various metals from the stream by scanning for x-ray absorption and firing precise puffs of air at the falling pieces to push them sideways into various sorting bins.

The remainder of the unsorted shredded material is known as fluff and contains mostly plastics, paper and other organic materials. When vehicles are shredded and processed in this manner for recycling, often a large mass of fluff results from the plastics used in the seat cushions, dashboard, roof liner, carpeting, and so forth. There are not many well-established processes for further separation and recycling of fluff, other than incineration or pyrolysis.

WASTE MANAGEMENT TECHNIQUES:

Managing municipal waste, industrial waste and commercial waste has traditionally consisted of collection, followed by disposal. Depending upon the type of waste and the area, a level of processing may follow collection. This processing may be to reduce the hazard of the waste, recover material for recycling, produce energy from the waste, or reduce it in volume for more efficient disposal.Collection methods vary widely between different countries and regions, and it would be impossible to describe them all. In Australia most urban domestic households have a 240 litre (63.4 gallon) bin that is emptied weekly by the local council. Many areas, especially those in less developed areas, do not have a formal waste-collection system in place.In Canadian urban centres curbside collection is the most common method of disposal, whereby the city collects waste, and or recyclables, and or organics on a scheduled basis from residential areas. In rural areas people dispose of their waste at transfer stations. Waste collected is then transported to a regional landfill.Disposal methods also vary widely. In Australia, the most common method of disposal of solid waste is in landfill sites, as it is a large country with a low-density population. By contrast, in Japan it is more common for waste to be incinerated, because the country is smaller and land is scarce.

Landfill

Disposing of waste in a landfill is the most traditional method of waste disposal, and it remains a common practice in most countries. Historically, landfills were often established in disused quarries, mining voids or borrow pits. A properly-designed and well-managed landfill can be a hygienic and relatively inexpensive method of disposing of waste materials in a way that minimizes their impact on the local environment. Older, poorly-designed or poorly-managed landfills can create a number of adverse environmental impacts such as wind-blown litter, attraction of vermin, and generation of leachate which can pollute groundwater and surface water. Another byproduct of landfills is landfill gas (mostly composed of methane and carbon dioxide), which is produced as organic waste breaks down anaerobically. This gas can create odor problems, kill surface vegetation, and is a greenhouse gas.

Design characteristics of a modern landfill include methods to contain leachate, such as clay or plastic lining material. Disposed waste is normally compacted to increase its density and stabilize the new landform, and covered to prevent attracting vermin (such as mice or rats) and reduce the amount of wind-blown litter. Many landfills also have a landfill gas extraction system installed after closure to extract the landfill gas generated by the decomposing waste materials. Gas is pumped out of the landfill using perforated pipes and flared off or burnt in a gas engine to generate electricity. Even flaring the gas is a better environmental outcome than allowing it to escape to the atmosphere, as this consumes the methane, which is a far more potent greenhouse gas than carbon dioxide.

Many local authorities, especially in urban areas, have found it difficult to establish new landfills due to opposition from owners of adjacent land. Few people want a landfill in their local neighbourhood. As a result, solid waste disposal in these areas has become more expensive as material must be transported further away for disposal (or managed by other methods)

This fact, as well as growing concern about the impacts of excessive materials consumption, has given rise to efforts to minimize the amount of waste sent to landfill in many areas. These efforts include taxing or levying waste sent to landfill, recycling the materials, converting material to energy, designing products that use less material, and legislation mandating that manufacturers become responsible for disposal costs of products or packaging. A related subject is that of industrial ecology, where the material flows between industries is studied. The by-products of one industry may be a useful commodity to another, leading to a reduced materials waste stream.

Some futurists have speculated that landfills may one day be mined: as some resources become more scarce, they will become valuable enough that it would be economical to 'mine' them from landfills where these materials were previously discarded as valueless. A related idea is the establishment of a 'mono-fill' landfill containing only one waste type (e.g. waste vehicle tires), as a method of long-term storage.

Incineration

Incineration is a waste disposal method that involves the combustion of waste at high temperatures. Incineration and other high temperature waste treatment systems are described as "thermal treatment". In effect, incineration of waste materials converts the waste into heat, gaseous emissions, and residual solid ash. Other types of thermal treatment include pyrolysis and gasification.

A waste-to-energy plant (WtE) is a modern term for an incinerator that burns wastes in high-efficiency furnace/boilers to produce steam and/or electricity and incorporates modern air pollution control systems and continuous emissions monitors. This type of incinerator is sometimes called an energy-from-waste (EfW) facility.Incineration is popular in countries such as Japan where land is a scarce resource, as they do not consume as much area as a landfill. Sweden has been a leader in using the energy generated from incineration over the past 20 years. Denmark also extensively uses waste-to-energy incineration in localized combined heat and power facilities supporting district heating schemes.

Incineration is carried out both on a small scale by individuals, and on a large scale by industry. It is recognized as a practical method of disposing of certain hazardous waste materials (such as biological medical waste), though it remains a controversial method of waste disposal in many places due to issues such as emission of gaseous pollutants.Breaking down complex chemical chains such as dioxin through the application of heat usually cannot be done by simply burning the material at the temperatures seen in an open-air fire. It is often necessary to supplement the combustion process with gas or oil burners and air blowers to raise the temperature high enough to result in molecular breakdown. Alternately, the exhaust gases from an natural air fire may pass through tubes heated to sufficiently high temperatures to trigger thermal breakdown.Thermal breakdown of pollutant molecules can indirectly create other pollution problems. Dioxin breakdown begins at 1000°C, but at the same time poisonous nitrogen oxides and ozone begin to form when atmospheric nitrogen and oxygen break down at 1600°C. This undesired oxide formation may require further catalytic treatment of the exhaust gases.

Composting and Anaerobic Digestion:

Waste materials that are organic in nature, such as plant material, food scraps, and paper products, are increasingly being recycled. These materials are put through a composting and/or digestion system to control the biological process to decompose the organic matter and kill pathogens. The resulting stabilized organic material is then recycled as mulch or compost for agricultural or landscaping purposes.

There are a large variety of composting and digestion methods and technologies, varying in complexity from simple windrow composting of shredded plant material, to automated enclosed-vessel digestion of mixed domestic waste. These methods of biological decomposition are differentiated as being aerobic in composting methods or anaerobic in digestion methods, although hybrids of the two methods also exist.Examples:The Green Bin Progammeme, a form of organic recycling used in Toronto, Ontario and surrounding municipalities including Markham, Ontario, Canada, makes use of anaerobic digestion to reduce the amount of garbage shipped to Michigan, in the United States. This is the newest facet of the 3-stream waste management system has been implemented in the city and is another step towards the goal of diverting 70% of current waste away from the landfills.

Green Bins allow any organic waste that in the past would have formed landfill waste to be composted and turned into nutrient rich soil. Examples of waste products for the Green Bin are food products and scraps, soiled papers and sanitary napkins.

The Green Bin Progammeme, a form of organic recycling used in Toronto, Ontario and surrounding municipalities including Markham, Ontario, Canada, makes use of anaerobic digestion to reduce the amount of garbage shipped to Michigan, in the United States. This is the newest facet of the 3-stream waste management system has been implemented in the city and is another step towards the goal of diverting 70% of current waste away from the landfills. Green Bins allow any organic waste that in the past would have formed landfill waste to be composted and turned into nutrient rich soil. Examples of waste products for the Green Bin are food products and scraps, soiled papers and sanitary napkins. Currently Markham, like the other municipalities in the

The Green Bin Progammeme is currently being studied by other Municipalities in the province of Ontario as a way of diverting waste away from the landfills. Notably, Toronto and Ottawa are in the preliminary stages of adopting a similar progammeme.The City of Edmonton, Alberta, Canada has adopted large-scale composting to deal with its urban waste. Its composting facility is the largest of its type in the world, representing 35 per cent of Canada's centralized composting capacity. The $100 million co-composter and various recycling progammemes enable Edmonton to recycle 60% of its residential waste. The co-composter itself is 38,690 square meters in size, equivalent to 8 football fields. It's designed to process 200,000 tons of residential solid waste per year and 22,500 dry tons of bio-solids, turning them into 80,000 tons of compost annually.

MECHANICAL BIOLOGICAL TREATMENT

Mechanical biological treatment (MBT) is a technology category for combinations of mechanical sorting and biological treatment of the organic fraction of municipal waste. MBT is also sometimes termed BMT- Biological Mechanical Treatment- however this simply refers to the order of processing.

The "mechanical" element is usually a bulk handling mechanical sorting stage. This either removes recyclable elements from a mixed waste stream (such as metals, plastics and glass) or processes it in a given way to produce a high calorific fuel given the term refuse derived fuel (RDF) that can be used in cement kilns or power plants. Systems which are configured to produce RDF include Herhofand Ecodeco. It is a common misconception that all MBT processes produce RDF. This is not the case. Some systems such as ArrowBio simply recover the recyclable elements of the waste in a form that can be sent for recycling.

The "biological" element refers to either anaerobic digestion or composting. Anaerobic digestion breaks down the biodegradable component of the waste to produce biogas and soil conditioner. The biogas can be used to generate renewable energy. More advanced processes such as the ArrowBio Process enable high rates of gas and green energy production without the production of RDF. This is facilitated by processing the waste in water. Biological can also refer to a composting stage. Here the organic component is treated with aerobic microorganisms. They break down the waste into carbon dioxide and compost. There is no green energy produced by systems simply employing composting.

MBT is gaining increased recognition in countries with changing waste management markets such as the UK and Australia where WSN Environmental Solutions has taken a leading role in developing MBT plants.

Pyrolysis and Gasification:

Pyrolysis and gasification are two related forms of thermal treatment where waste materials are heated to high temperatures with limited oxygen availability. The process typically occurs in a sealed vessel under high pressure. Converting material to energy this way is more efficient than direct incineration, with more energy able to be recovered and used.

Pyrolysis of solid waste converts the material into solid, liquid and gas products. The liquid oil and gas can be burnt to produce energy or refined into other products. The solid residue (char) can be further refined into products such as activated carbon.

Gasification is used to convert organic materials directly into a synthetic gas composed of carbon monoxide and hydrogen. The gas is then burnt to produce electricity and steam. Gasification is used in biomass power stations to produce renewable energy and heat.

THE WASTE PROBLEM IN THE INDUSTRIAL SOCIETY

The waste management system has been defined as the collection of economic activities that accepts solid or liquid waste inputs from households, and/or industry and either recycles or disposes of them. This has been called an "industry", and (in recent years) even a "growth industry". There are a number of large national and some multinational firms that operate profitably in this field. But it is important to recognize from the outset that, while growth in this sector adds to GDP, it is not an indicator of increasing human welfare.

Market Failure

The need for a separate waste management system outside and distinct from both the productive part of the economy (the "factory") and the consumption unit (the "household") is a reflection of market failure in the sense that neither producers nor consumers normally pay for waste management services for their own sake. The fact that households normally pay a small charge for refuse collection is a minor exception to this rule. Despite lip-service to environmental concerns, most waste generators still act and think as though the environment is a free waste-disposal sink.

The fact that rivers and lakes, landfills and even the atmosphere and the oceans are finite, and are becoming over-used, has not been translated into a market-determined scarcity cost for environmental services, as would normally be the case for a scarce resource. The reason this has not happened is that the resources in question – except for landfills – are common property "public goods" that cannot be monopolized or "owned" by any individual. This is the essential market-failure.

To dispose of wastes without overloading the environment, it has been necessary to invent a "waste management" sector to provide waste disposal services. In the case of households and communities, these services are paid for by public sector agencies, which are reimbursed through refuse collection fees, water charges and/or property taxes. In the case of industry some waste management services are paid directly (by industry) to comply with regulations imposed and enforced by the public authorities. There is only a weak and imperfect link between the quantity and type of waste generated and the charges for treatment and disposal. In many cases, the costs attributed to polluters are too low to induce significant changes in their behaviour. In many cases, too, the level of treatment or amelioration is inadequate and the resulting environmental damages are accumulating.

In short, there is no competitive free market between buyers and sellers of environmental services, *as such*. Instead, the benefits of waste management services are a rather diffuse common good. This good is enjoyed equally by all, but it is not paid for in proportion by those who generate the wastes. The intervention by public authorities is an inefficient "second best" solution, relying more on rules and regulations and less on economic in centives than would be optimal.

Statistics of the waste management "sector" must be treated with care; they can easily be misinterpreted. (For instance, statistics showing an increase in the aggregate quantity of solid and liquid wastes treated may or may not imply that the quantity of wastes being generated is growing. By the same token, they may or may not imply that the overall pressure on the environment is being reduced.

Such a trend may simply be a measure of increasing regulation. Or it could reflect increasing separation between the various stages of production and processing, and of consumption, i.e. consequences of specialization, centralization and scale). So, although statistics of the waste management sector have value, on their own they are not likely to be optimal measures of distance from or approach to long-term sustainability.

Delimitation of the sector in the SIP project and overlaps with the other SIPs

We have already defined the waste management system as the collection of economic activities that receives solid or liquid waste inputs from households, industry and/or other economic sectors and either recycles or disposes of them. For statistical purposes it is useful to represent the sector as any other industrial system: a set of processes characterized by an input - output structure. In this specific case, the process inputs are the different wastes to be treated, whereas the outputs are constituted by the relatively inert residuals of the treatment processes, air and water emissions.

In order to provide an homogeneous representation, a similar input - output structure should be applied to all the six economic sectors of the SIP project, including Tourism and Transport, even if the latter do not produce physical products as outputs. Such a difficulty may be overcome simply by considering that these two sectors provide a service instead of a physical good.

The general view of the economic activities of each SIP that results from this general input - output representatio. The figure also puts a special focus on the major waste streams produced by these sectors and accepted by the waste management system.

Given the complex interlinks occurring between the waste management system and the other economic sectors, it is important to define appropriate system boundaries among the sectors. However, the identification of these boundaries is strongly dependent on how waste management is defined.

We have adopted the definition of waste management as a sector providing services of waste collection, treatment and disposal for the various economic activities and municipalities. However, depending on the industrial structure in Member

States, the management of waste may also take place as a sub-activity of each economic activity. Thus, waste management could also be regarded as an industrial operation or a *functional unit* occurring in all the economic activities. This would also be in accordance with the concept of statistical unit Functional. units such as waste treatment facilities in an establishment or an enterprise can be considered as a Unit of Homogenous Production. Indeed, the functional role rather than the sectoral one may be the most important for the overall waste management sector in a Member State.

The issue is particularly important for the case of industrial waste, because a large number of manufacturing industries provide on-site treatments before discarding their residuals for final disposal. Depending on the perspective adopted (waste management as an independent sector or as a functional unit), the environmental pressure resulting from these on-site treatments could be allocated either to industry (where the treatments take place) or to the waste management sector (because, after all, even on-site treatments are operations of "waste management").

If we adopted a classification system based on the second principle (aggregation of homogeneous processes), the pressure of the on-site treatments should be allocated to the waste management sector.

However, for purposes of statistical accounting such an approach would present some problems of data collection. It is more difficult to gather precise data on the intermediate products of a production unit than to obtain overall input - output Therefore, it seems more convenient to adopt the *mill gate* as a conceptual boundary between the two sectors and to allocate to the primary waste generators the pressure of the on-site treatments. The waste management sector will be accounted only for the environmental pressure generated by the treatment and final disposal of wastes which leave the mill gate of primary waste generators. That is, those wastes *not* further treated or recycled by industry and consumers.

System Boundaries between Industry and Waste Management

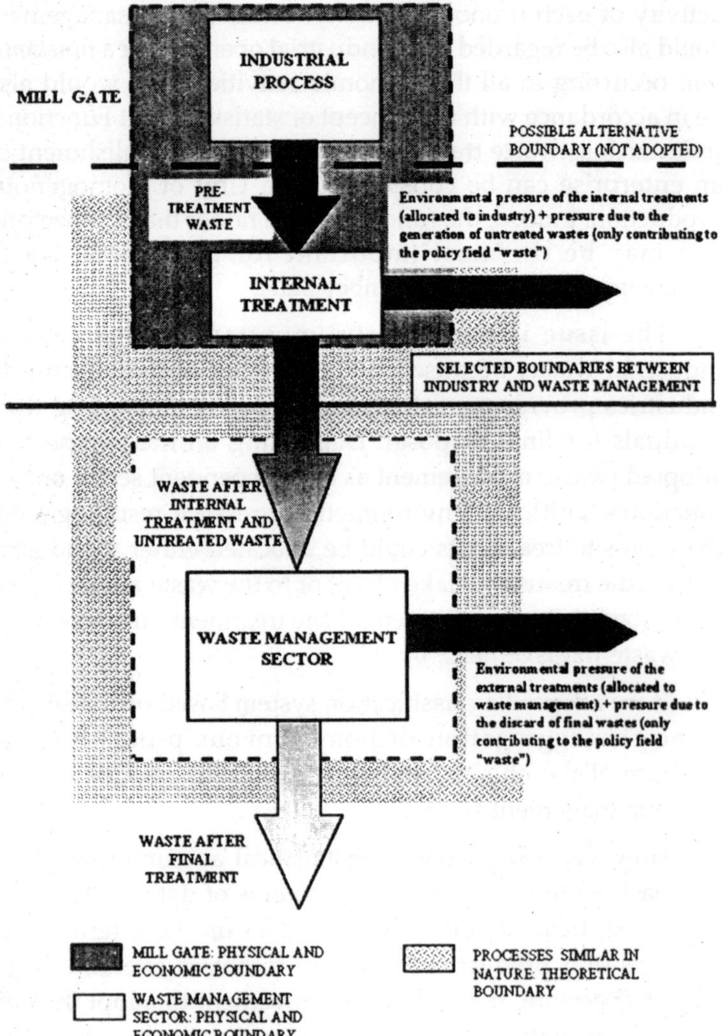

Fig.: System Boundaries between Industry and Waste Management

As a consequence, the manufacturing sector will be accounted both for the pressure generated by on-site treatments of primary wastes (in principle contributing to all the policy fields except "waste") and for the pressure related to the generation of wastes which leave the mill gate (both treated and untreated). It is only the latter pressure that contributes to the policy field "waste".

On the other hand the waste management sector will be accounted for the pressure due to off-site treatments (again contributing to all the policy fields except "waste") and for the pressure of final wastes to be disposed of (only contributing to the policy field "waste").

The Waste management sector and the environmental policy fields

In principle, the waste management sector produces environmental pressure on the ten policy fields according to the theoretical scheme presented. A presentation of the specific environmental pressure caused by the sector is presented. In the following some general aspects are summarized relating to the major environmental concerns and their policy implications in the context of waste management.

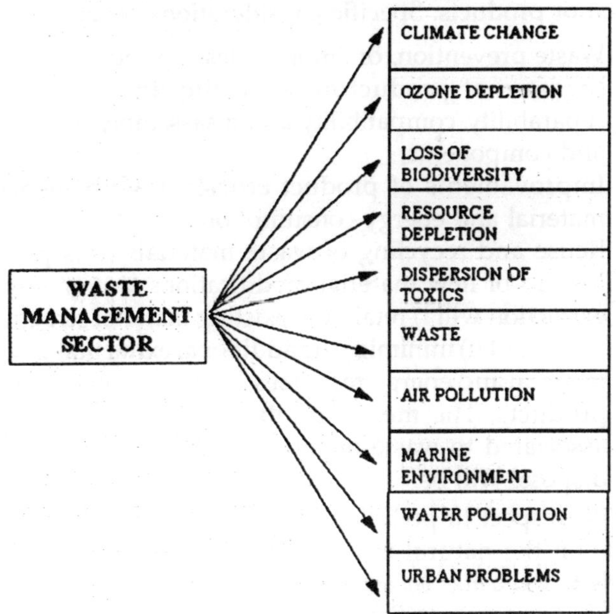

Fig. The waste management sector and the ten policy fields

Being a collection of many different pressures arising from great number of waste materials and waste management systems or processes that accepts, treats and disposes of a large number of substances of a various nature, the pressures of the waste management sector may be generally described by means of pressure characterization and a twofold representation of underlying factors. A first order description of both the different waste streams and waste management systems. A general view of the major material and energy flows generated by the various economic activities and connected to the sector is also presented

PRESSURE CHARACTERISTICS

As a general remark, major environmental concerns about waste generation relate both to the problem of resource depletion and to that of environmental pollution

Resource aspects include material efficiency and resource intensity for the main production processes from extraction of natural resources, manufacturing, to consumption and final disposal of products. Specific considerations cover:

- Waste prevention, or "more of less" principle;
- Increase of product/material life time, increase of separability, compatibility and disassembly of materials and components;
- Improvements of product efficiency with respect to material and energy consumption;
- Reuse and recycling of waste materials (this practice, instead of new material modifications and/or resource extraction will i) retain the existing material structure of waste and ii) minimize the additional extraction of virgin material and energy resources for the production of new products. The more cumulative resource inputs are associated to waste material the less sustainable their disposal will be.

With respect to pollution, the waste management sector impacts on the natural environment by *discarding treatment residuals to air, water bodies or soil.* As a consequence, several objectives can be identified such as:

- To close material cycles; minimization of loading
- To prevent and minimize generation of hazardous waste

- To manage waste collection, handling, treatment and final disposal
- To increase reuse and recycling in order to minimize emissions from waste incineration installations
- Proximity principle: to limit transportation of waste
- To control environmental pollution preferably at the sources of waste

The long-term objectives of the EU waste policies is to conserve raw materials and energy as well as to ensure the high level of environmental protection. The cause-effect relationships of these two types of pressures are not always recognized. However the evaluation of the overall environmental impacts of waste already includes quantitative (the amount of waste generated), and qualitative aspects (their hazardousness).

Major Waste Streams

An example of the large number of different waste streams produced by the most important economical activities is presented, together with some additional information regarding possible treatment processes

Management Systems

In principle, we note four major economic processes, plus an additional external non-economic sector to include all the substances which are directly discharged into the environment without any special treatment (mainly agriculture residuals). Thus, the processes for the management of solid and liquid wastes can be subdivided as follows:

- Landfilling or land application
- Incineration
- Treatment of liquid wastes, including sewage
- Treatment of special and hazardous wastes
- Direct disposal without treatment

The solid components of materials treated in the second, third and fourth sub sectors usually ends up in landfills for a definitive final disposal. Thus the soil can be regarded as the final "sink" for most wastes (except for the soluble residuals discharged to water courses and the airborne emissions from incineration).

MAIN PRESSURES BY POLICY FIELD

Air Pollution

In the policy field of air pollution all the environmental effects caused at a local up to regional level by the various types of airborne emissions to be included. The gaseous substances that are responsible only for global environmental effects, such as ozone layer depletion or climate change have not been included in the policy field Air Pollution, but in the policy field "Ozone Layer Depletion" and "Climate Change" respectively. For instance the group of CFCs has not to be considered in this particular policy field. In fact, these compounds are usually accounted as airborne emissions, but none of them has a local specific polluting effect.

Carbon monoxide, nitrogen oxides have specific effects on human health, such as particulate and volatile organic compounds. Sulfur oxides are responsible for the problem of acidification, that has also a global aspect. However, given that which is proposed in the guidelines suggested by DG XI, their contribution is included in this policy field. Methane has not been included in the list because its contribution to local air pollution problems is insignificant.

The contribution of the waste management sector to this area is twofold: treatment operations per se and transportation by trucks

Concerning the contribution related to the treatment operations, the only process that is relevant for air pollution is incineration. However, the amount of emissions generated is strongly dependent on the type of waste incinerated The contribution due to the transportation system adopted is also easily identifiable. In fact the large majority of solid wastes are transported by truck. Thus, standard emissions factor can be applied to estimate the amount of atmospheric emissions that have to be allocated to the waste management sector. Anyway, this is still not possible and this sort of data are included under road transport statistics.

Climate Change

The potential problem of climate change can be correlated to that of enhancement of the greenhouse effect. This phenomenon is caused by gaseous substances that absorb and re-radiate the infra-red radiation (IR) from the earth surface, instead of allowing this IR radiation to escape into outer space. In consequence, a progressive enhancement of the average temperature of the planet is expected and a complex set of consequent climate changes would follow.

Given the attention that the problem of greenhouse effect has received by the international community in recent years, a well-established set of pressure indicators is available to provide a qualitative first-order assessment of this environmental theme. A complete list of substances that contribute to this problem together with their relative global warming potential can be found, for instance, The most important gaseous emission for this specific problem is the carbon dioxide. In addition, a global warming potential has been identified for a large number of other substances that may present a potential contribution to the greenhouse effect. The most important contributors are: methane, nitrous oxide (N_2O), several chlorofluorocarbons (CFCs), and other chlorinated compounds such as CCl_4. However, given the relatively small or even negligible contribution of the waste management sector to the anthropogenic emissions of some of these compounds, it is possible to circumscribe the assessment to a limited number of substances:

- CO_2 is mainly emitted by incinerators and old style landfills. An additional contribution to the production of CO_2 has also to be allocated to the transportation systems adopted for moving wastes from the generation site to the management site.
- CH_4 is produced in landfills by anaerobic digestion of organic matter. Some modern landfills are equipped with

special systems for gas extraction and reuse. However, in most sites, methane is directly emitted to the atmosphere.

- these substances were mainly used as refrigerants for their characteristics of thermal capacity and low cost, as propellants in many common aerosol sprays, and for the production of plastic foams used for thermal insulation. Though production has been phased out in Europe, some refrigerants and foamed plastics are still in use. Some of these are still disposed of in open landfills.

To summarize, a consistent evaluation of the contribution of the waste management sector to the policy field of climate change can be provided when information is given about the atmospheric emissions of CH_4, CO_2

Loss of Biodiversity

No major direct contribution of the waste management sector to the problem of Loss of Biodiversity was identified.

Marine Environment and Coastal Zones

The specific impact of the waste management sector on the marine environment arises almost entirely from water pollution.

Ozone Layer Depletion

As for the case of climate change, the problem of the ozone layer depletion is well studied. Indeed, it was predicted theoretically before it was first observed. The progressive depletion of the stratospheric ozone layer is caused by some specific compounds that do not break down in the lower atmosphere, but migrate to the stratosphere, where they are broken up by UV radiation releasing chlorine atoms in the stratosphere. Chlorine atoms interact with the ozone molecules starting a chain reaction that progressively destroys ozone (O_3), which is the natural shield for the dangerous UV radiation.

A specific group of inert and very stable chemical compounds is largely responsible for the depletion of the ozone layer: the chlorofluorocarbons (compounds containing atoms of chlorine, fluorine and carbon, but no hydrogen). In addition, some other compounds containing chlorine or bromine, such as CCl_4 and CF_3Br are characterized by a relatively high ODP. However, a quantification of emissions suffice for purposes of tracking the contribution of the waste management sector. In fact emissions of other attributable to the waste management sector are probably negligible.

The major contribution of the waste management sector to this problem is from the containing devices disposed of in open landfills.

Resource Depletion

The waste management sector is both a "sink" and a "source" for materials and energy. Landfills are obviously sinks, but some resource recovery and recycling also occurs in municipal waste operations. Glass bottles, aluminum cans and paper are among the items being recycled, and also more and more incinerators are generating electric power as a by-product.

The contribution of the waste management sector to resource depletion is mostly represented by the consumption of energy resources (mainly electricity) for the treatment processes or for waste transportation operations. The latter example plays an important role, as large quantities of solid waste are moved by truck from the generating site to the treatment site. On the other hand, incineration is a special case, where energy resources are also produced: in fact sometimes wastes are used as a fuel for the generation of heat or electrical energy.

Dispersion of Toxic Substances

All the environmental problems related to the substances that are likely to cause serious ecotoxicological problems or human diseases, especially when introduced in the food chain are included in this policy field.

As a first consequence, the contribution to this policy field should be represented by a very large number of chemical compounds contained in the various wastes generated, especially by chemical industries. In fact, several hundreds of chemical compounds are listed as potentially toxic by different sources. However, the treatment systems adopted for the majority of hazardous wastes are capable of neutralizing almost completely the potential danger of these substances. Therefore, paradoxically, we could state that toxicity does not represent a real problem for the most part of toxic wastes.

On the other hand, an important environmental concern may be represented by all the wastes that are not classified as hazardous but that may contain some amounts of toxics. In fact, these wastes do not receive special treatments, and are generally disposed in open landfills. The progressive action of atmospheric agents, such as rain or wind, may contribute to releasing the toxic components embodied in the waste material. If landfills are not properly managed, i.e. equipped with liners, these substances are more likely to migrate under the form of leachate and to pollute soil and groundwater, thus affecting animal and human food chains.

Urban Environmental Problems

The definition of "Urban problems" is very broad and several effects could be included. It is proposed to consider only the problems of traffic, noise and odours as they are specifically related to urban areas. In addition, we propose to include the problem of land use, as it represents an important issue for the waste management sector. In some environmental statistics this theme is usually considered separately in a specific category. However, given the fact that none of the identified policy fields seems to include it, we suggest to quantify the related environmental pressure at this stage of the analysis.

Other phenomena that may contribute to the enhancement of the environmental pressure in urban zones have already been considered in other policy fields (air pollution, for instance).

A number of environmental effects related to the urban problems may be allocated to the sector of waste management at different levels. All of them are attributable to the presence of waste treatment sites, that is landfills or incinerators, in urban areas.

A first contribution is represented by the intensive movement of trucks around the management sites, with specific effects on air pollution, congestion of traffic, and noise. A second contribution is specifically related to the management systems adopted. In fact, both landfills and incinerators cause the emission of malodorous substances and present problems of noise generation. Unfortunately, it is quite difficult to provide a quantitative assessment of noise and odor levels. In fact, this objective could be achieved only by means of precise on-site measures of the noise sources and malodorous emissions. The latter approach, though feasible in the realm of product-oriented LCA, is not easily applicable for statistical purposes and should be replaced by a simpler methodology.

Waste

"Waste" is both a policy field and an economic sector. Here we refer only to the policy aspect. Given the broad and various meanings associated to the concept of "wastes", we propose to consider it in its broader and common meaning. A waste is a useless, space-consuming, generally non-hazardous solid or liquid material that must be discarded and permanently stored in a specific site if not reused or recycled. As a result, the environmental pressure associated to this policy field is to be related to the amount of space occupied by the final solid residuals of various management operations, including waste treatment sequences.

The relative contribution of the waste management sector to this policy field is important and diversified. In a simplified way, three general categories of final solid wastes may be identified, namely:

- Non-hazardous solid wastes: this category is represented by all those solid wastes that are directly disposed of without special treatments or transformation processes. A number of different waste streams are included in this group, e.g.: residuals from agricultural activities and feedlots, residuals from crop or wood harvesting, municipal solid wastes, construction debris, old machines or components directly disposed of in landfills;

- Industrial non-hazardous solid wastes: this broad category includes all the industrial solid wastes that can be collected, stored and disposed of without any special treatments for neutralising them. As examples, we may mention all the wastes from extraction and beneficiation processes, residuals from particulate emission control systems, different sludge from metal industries;
- Solid residuals from treatment of hazardous or special wastes: this group is mainly constituted by the solid residuals that occur from liquid waste, sewage or hazardous waste management. In fact, these waste streams, often in a liquid state, usually receive special treatments to neutralize dangerous components and to separate them from the cleaned effluents. These components, reduced to a relatively inert form such as concrete blocks, are finally discarded as non-hazardous solid wastes. Many different waste categories receive these type of treatment, thus generating a solid inert residual. The most important are: sewage, pulping sludge from pulp and paper industries, process residuals from chemical industries, liquid effluents from almost all the manufacturing industries, drilling fluids from oil extraction.

Water Pollution and Water Resources

It is proposed to limit the assessment of the problem of water pollution to the substances that have a local polluting effect on the environment, but not specific toxic consequences, since the latter have already been included under the item "Dispersion of Toxic Substances".

In the case of waste management, COD, BOD and TSS are found in the liquid effluents that are released to water courses, both directly and after specific treatment. In fact, even if treatment systems are adopted, they may not be able to completely remove the whole range of pollutants contained in the effluents.

The problem of water pollution is specifically related to the treatment of liquid wastes, such as sewage, process and cooling water from chemical and metallurgical industries, and some liquid residuals of various types from different manufacturing industries. The problem is particularly important with regard to the pulp and paper industry. In fact the large amount of organic substances processed in this sector causes high levels of BOD and TSS in the liquid effluents. These are difficult to eliminate completely via end-of-pipe emission control systems.

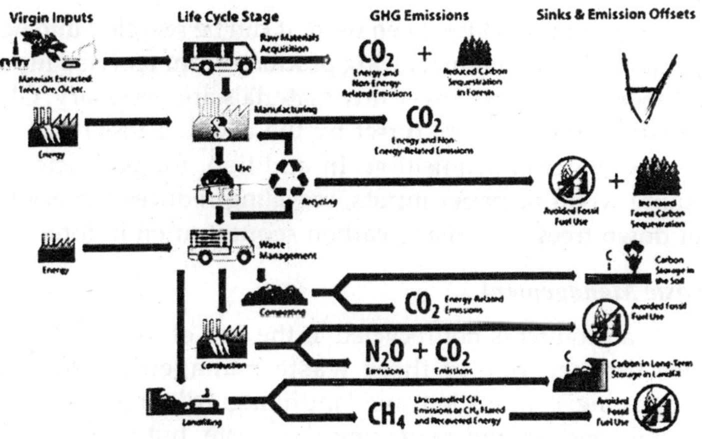

Fig.: Life-Cycle of Waste Image

The four main stages of product life-cycles, all of which provide opportunities for GHG emissions and/or offsets. These stages are: raw material acquisition, manufacturing, recycling, and waste management.

Raw Material Acquisition

All products use inputs of raw materials, such as metal ore, petroleum, trees, etc. Extracting and transporting these materials entails the combustion of fossil fuels for energy, which results in emissions of carbon dioxide. These fossil fuels must be extracted themselves, which requires additional energy use.

Manufacture

The processes that transform raw materials into products require the combustion of fossil fuels for energy. Again, energy use produces GHG emissions both directly from the combustion of fossil fuels (mainly in the form of carbon dioxide) and from the upstream energy used to obtain and transport those fossil fuels. In addition, some manufacturing processes release other GHGs, although the type and amount of these emissions are specific to the manufacturing processes for each material.

Recycling

Once a product has been used, it can be recycled into new products. While manufacturing products from recycled inputs still requires energy, fewer raw materials are necessary. GHG emissions are therefore offset by the avoided fossil fuel use for raw material acquisition. In addition, for products that require wood or paper inputs, recycling reduces the need to cut down trees, increasing carbon sequestration in forests.

Waste Management

If a product is not recycled at the end of its useful life, it goes through one of three waste management options: composting, combustion, and landfilling. All three use energy for transporting and managing the waste, but they produce additional GHGs to varying degrees.

Composting – an option for organic materials such as food scraps and yard waste – releases some non-biogenic carbon dioxide associated with transporting and turning the compost. However, some of the carbon contained in organic materials is returned and stored in the soil and therefore not released into the atmosphere.Combustion releases both carbon dioxide and nitrous oxide (a GHG that is 310 times more potent that carbon dioxide). However, some of the energy released during combustion can be harnessed and used to power other processes, which results in offset GHG emissions from avoided fossil fuel use.Landfilling, the most common waste management practice, results in the release of methane from the anaerobic decomposition of organic materials. Methane is 21 times more potent a GHG than carbon dioxide. However, landfill methane is also a source of energy, and some landfills capture and use it for energy. In addition, many materials in landfills do not decompose fully, and the carbon that remains is sequestered in the landfill and not released into the atmosphere.

RADIOACTIVE WASTE MANAGEMENT

Framing the Problem

The United States is at a "gridlock" position regarding nuclear waste management. Existing nuclear power plants, left to manage wastes in the absence of a coherent national policy, have become de facto long-term storage sites, using facilities designed only to temporarily house such materials. Radioactive waste has emerged as one of the issues inhibiting further development of the nuclear power industry, and the safety implications of forcing every power plant to handle wastes on a longer-term basis are severe.Former nuclear weapons production sites face even more significant problems with radioactive waste management. The scale and scope of the cleanup at these sites is enormous; officials estimate that seventy-five years and $300 billion will be required to remediate these facilities. The Department of Energy (DOE), which is responsible for these sites, faces both an environmental and an administrative quagmire as it attempts to clean up after fifty years of nuclear weapons production.The lesson that follows is an introduction to these issues, focusing on an understanding of how, over time, the problems associated with radioactive waste have developed. This lesson is predicated on the assumption that radioactive waste management is not a single task to be accomplished but is rather a multidimensional issue that needs to be understood and addressed physically, socially, and historically.

The multiple aspects of radioactive waste management this lesson examines include physical and value-oriented issues. What radioactive waste is, where it is, how it has been generated and handled, what some of the difficulties are in handling, treating, and disposing of the stuff—these are central issues in radioactive waste managment. Values are also critical in understanding radioactive waste, and must be understood in their social and historical contexts—how radioactive waste has been understood (or misunderstood, as the case may be), how it has meant different things to different people, and how these meanings have changed over time. An understanding of these value-oriented issues will enable a discussion of the political problem of radioactive waste, how it has developed into a metaphorical "hot potato" that nobody is willing to hold. Understand both values and the physical aspects of radioactive waste will help foster an understanding of the bureaucratic problems involved in managing this material —resolving the radioactive waste problem will involve administrative and managerial feats that the DOE has not yet shown itself capable of handling.

The radioactive waste "problem" is not a new one; health and safety concerns have been associated with radioactive materials throughout the twentieth century. Indeed, such concerns have existed almost from the discovery of radioactivity in 1895, when a German physicist, Willem Roentgen, identified what he called "x-rays," and developed a technique for producing them. The medical community quickly adopted x-rays as a useful diagnostic tool; within a year, x-ray machines could be found in every major city. In early 1890's identified a new element, radium, that had radioactive properties. Radium, like x-rays, was adopted for several industrial uses.

Observers soon noted that these radioactive materials were associated with harmful side effects. Medical x-ray workers and many scientists were suffering from skin burns, blood disorders, and a variety of otherwise rare cancers; indeed, Marie Curie suffered terribly from bone diseases stemming from her prolonged exposure to radium and other radioactive materials. The women who worked as watch-dial painters began to develop horrible forms of jaw and throat problems. Health workers learned that these women would use their mouths to "point" their brushes, thereby ingesting what often amounted to lethal quantities of radioactive compounds.Several interested organizations and individuals moved, in response to the recognition of the health effects of radioactive materials, to form independent bodies to study the issues and set voluntary standards for exposure.

The Manhattan Project: Large-Scale Work with Radioactive Materials

Industrial work with radioactive materials was relatively small in scale through the 1930s, but events during and after World War II changed matters dramatically. Indeed, the work done to construct atomic weapons, begun as war raged across Europe and the Pacific, continues to have profound environmental and political effects in the contemporary world.

A. The Decision to Build the Bomb

Efforts to construct an atomic bomb in the United States began in the late 1930s, under the assumption that the Allies were in a race with Germany. In 1938, two German physicists announced that they had demonstrated fission—the splitting of the atom. Scientists recognized that a controlled fission reaction could be made into a new, incredibly destructive weapon. Many of the most eminent physicists in the. were emigres from Germany, Austria, Hungary, and Italy; having experienced fascism, they were alarmed by the prospect of an atomic weapon in the hands of Hitler's regime. The efforts of these emigre scientists were crucial in prompting the American government to undertake what became known as the Manhattan Project.The Manhattan Project refers to the efforts of the Manhattan Engineering District (MED), the organization created by the U.S. Army to carry out the atomic bomb development progammeme. This was truly an enormous industrial and scientific undertaking, and was the largest construction project ever undertaken at the time.

Health and Safety in the Manhattan Engineering District

Radiation safety was a priority in the MED. Personnel protection was strongly emphasized, even more than in other types of production work, and the NCRP safety guidelines proved quite effective in this regard. Scientists were in short supply and even enlisted men (who died by the thousands overseas) were well protected. The unique nature of the research and production work made trained personnel even more valuable than normal; there simply were no replacements available. Management thus emphasized health and safety precautions, as the project could not afford to lose trained personnel to radiation exposure. Site selection reflected safety concerns of a different sort. The MED chose to locate its primary plutonium production facility (Hanford) in eastern Washington, rather than nearer to the labouratories in Chicago or Oak Ridge, because of fears of the possible consequences of a major accident. The Hanford facility utilized new and untested production techniques with extremely dangerous substances. Managers did not know exactly what the risks of operating such a plant were, but they assumed that they were significant. Thus, they chose to place the facility in a remote location so that if a catastrophic accident did occur it would affect relatively few people. Waste management was not a high priority during the war; this is not a surprise given the pressures of defeating Germany and Japan.

Although managers clearly understood that production processes would generate vast quantities of highly toxic materials, especially from the separations processes at Hanford, they did not give much thought to the development of a long-term waste disposal plan. The MED sought to minimize the immediate risks associated with these wastes, but deliberately deferred developing disposal solutions for the various waste management problems until after the war. Even though waste management and health concerns were re-evaluated after the war ended, the MED never made the solution of waste problems a priority. As it turned out, neither did its successor, the Atomic Energy Commission.

The Legacy of the MED and the rise of the Cold War

The MED was ultimately successful in developing atomic bombs, two of which were dropped on Japan in August, 1945. Atomic weapons became a central element of American diplomacy from the moment of their first use, and "the bomb" played a critical role in the emerging Cold War between the US and the USSR. The tensions between East and West emerged even before the war ended and escalated throughout the 1940s and 1950s. As a result, with the exception of a brief post-war lull, the American nuclear weapons production effort begun during the war continued at a high level well into the 1960s.The rapid transition from WWII to the Cold War helps to explain the lack of attention paid by the Manhattan Engineering District and the Atomic Energy Commission to waste managment issues. As atomic weapons became a defining symbolic and literal expression of American power, the need to continue production at all costs continued. There was no extended post-war shut-down of the wartime production network, no time to develop new approaches to waste production and management. The pressure of the Cold War made for a very uneasy peace indeed.

Index

F

G

H

I